Factorization of Boundary Value Problems Using the Invariant Embedding Method

In tribute to J.-L. Lions, master and inspirer.

Series Editor
Jacques Blum

Factorization of Boundary Value Problems Using the Invariant Embedding Method

Jacques Henry
Angel M Ramos

ISTE PRESS

ELSEVIER

First published 2016 in Great Britain and the United States by ISTE Press Ltd and Elsevier Ltd

ISTE Press Ltd
27-37 St George's Road
London SW19 4EU
UK

www.iste.co.uk

Elsevier Ltd
The Boulevard, Langford Lane
Kidlington, Oxford, OX5 1GB
UK

www.elsevier.com

British Library Cataloguing-in-Publication Data
A CIP record for this book is available from the British Library
Library of Congress Cataloging in Publication Data
A catalog record for this book is available from the Library of Congress
ISBN 978-1-78548-143-7

Contents

Preface

The invariant embedding method is archetypical of the scientific approach. In a first analytic phase, it consists of studying elementary parts of a given problem. Then, the study of the relations between these parts gives rise to a second synthetic phase that allows us to access the solution of the problem. This method is well known in control theory, where it was first introduced by R. Bellman. In the context of control theory, t denotes the time, u the state of the given system and v the control. The method is performed by embedding an optimal control problem, defined on a temporal horizon $[0, T]$ (where T can be finite or infinite), into a family of similar problems defined on $[t, T], 0 \leq t \leq T$, while additionally choosing an arbitrary initial state u_t at time t. Let t denote the present moment and u_t the current measured state. Invariant embedding consists of considering two neighboring time points t and $t + \delta t$ and requiring that the evolution of the state satisfies the state equation between the two time points with the restriction that their paths must be optimal on $[t, T]$ and on $[t + \delta t, T]$. We can then write an equation for the optimal value of the criteria as a function of t that will give us not only the optimal control for the simple case of the "open loop" $v^{\star}(t)$, which minimizes the objective functional assuming that the state satisfies the state equation exactly at every moment, but also the optimal control for the "closed loop" $v^{\star}(t, u_t)$, which yields the best control to be applied to the system at time t when its current state u_t is known, and assuming that it will not be perturbed in the future. Whenever computationally feasible, automation specialists prefer the closed-loop control because it is much more stable. This method was developed by R. Bellman and his students (see, in particular, [BEL 57a]).

In this publication, we will revisit the principle of the invariant embedding method and develop the idea further by applying it spatially to solve elliptic linear boundary value problems: starting from a problem defined on one domain Ω, we will define a family of similar problems on the subdomains $\Omega_s \subset \Omega$ indexed by a parameter s with values ranging from 0 to 1, such that $\Omega_s \subset \Omega_{s'}$ for $s < s'$ and $\Omega_1 = \Omega$. The set Ω_0 has null measure and may or may not be contained in the boundary Γ of Ω. The "mobile" boundary Γ_s of Ω_s, meaning the part of the boundary of Ω_s that does not belong to the boundary of Ω, sweeps the entire domain Ω as s varies from 0 to 1. To ensure that the sub-problems are well defined, we must impose a boundary condition on Γ_s. The value of this condition is arbitrary as is the case in control theory, where an arbitrary present state is chosen for invariant embedding. In the framework of the optimal control of linear systems with quadratic cost, invariant embedding can be used to define a linear operator connecting the present state to the optimal adjoint state at the same time. This operator satisfies a Riccati equation. In our approach, we introduce a similar operator connecting the boundary condition on the mobile boundary Γ_s to other, complementary conditions, typically the Dirichlet and Neumann conditions. This operator also satisfies a Riccati equation. For optimal control, the invariant embedding process results in a Riccati equation for the state-adjoint state operator that depends only on the nature of the control problem, and an evolution equation on a residual taking into account the particular features of the problem. The optimal control is then obtained, given the measurement of the present state of the system, by an affine relationship involving this operator and this residual. Likewise, with our invariant embedding approach for elliptic linear boundary value problems we can transform the problem for which the solution at each point of the domain depends on the set of all given data (right-hand side in the entire domain and boundary conditions on the whole boundary) into two decoupled parabolic-type Cauchy problems: one on the residual to be solved in the direction of the embedding and the other to be solved in the backward direction giving the solution to the boundary value problem. These problems are dependent on the given data of the boundary value problem. They use a Dirichlet-to-Neumann-type operator that satisfies a Riccati equation and is independent of the given data, depending only on the operator of the elliptic problem, on the geometry of the problem and on the definition of the invariant embedding. This transformation of the initial boundary value problem into two Cauchy problems that we can then solve successively is referred to as the "factorization of the boundary value problem". This is not a factorization of the elliptic operator, but of the boundary value problem itself, because the factorized problem satisfies the boundary conditions.

Our research began with the article [HEN 96], published in collaboration with J.-P. Yvon, in which we studied the control of an elliptic boundary value problem by a boundary condition. This problem inspired us to implement spatial invariant embedding in conjunction with the techniques of invariant embedding introduced in optimal control. The idea of spatial invariant embedding had already been used by R. Bellman and his collaborators for elliptic boundary value problems, mostly in rectangular domains [ANG 71a]. These studies were developed in a series of papers by E. Angel and diverse collaborators. In [ANG 68a], invariant embedding is applied to a boundary value problem for Laplace's equation in a rectangle and then extended to less regular regions. The problem is discretized by finite difference methods. We will develop this perspective in Chapter 5. In [ANG 69], he gives a method for constructing the solution to a boundary value problem in a composite domain in the case where the solutions are known in the subdomains. The paper [ANG 68b] presents a comparison between invariant embedding applied to a problem that is completely discretized by finite differences and continuous invariant embedding applied to a problem that is discretized only with respect to the transversal variable. In [ANG 70b], together with R. Kalaba, he studies the possibility of a factorization with only one sweep. In [ANG 70a], a link is established between a system of two (discretized) evolution equations, coupled and to be integrated in opposite directions, and a boundary value problem written in the form of a two-point boundary value problem. The paper [COL 71] presents a numerical method called diagonal decomposition, enabling the use of an iterative solution of Riccati equation using matrices of reduced size. The numerical aspects are revisited in [ANG 71b]. The idea of combining the direct numerical method resulting from the invariant embedding with an iterative method has been suggested. A description of a method for solving weakly nonlinear elliptic problems by invariant embedding in arbitrary 2D domains is also given in [ANG 86]. Nonlinearity is handled using the quasi-linearization method. Finally, [ANG 94] displays the use of a factorization method in the framework of parallel computing. Going back to earlier research, we find the study by Hadamard [HAD 10] on the variation of the Green's function of a boundary problem due to boundary displacement. This study introduces a differential equation with quadratic terms very similar to the Riccati equation in our approach. It does not seem, however, that the method of the spatial invariant embedding has been extensively studied from a theoretical point of view.

In this book, we explore the domain in which this method can be applied. For simplicity of presentation, we propose a particularly simple "model problem": a boundary value problem with the Laplace operator and a cylindrical domain. This cylindrical geometry imitates the control situation of an evolution problem (cylinder with respect to time) and allows a simple and natural embedding of the invariant with respect to the axis of the cylinder. Furthermore, in this model problem, we choose boundary conditions that simplify the determination of the initial conditions of the factorized problem. The very simple formal derivation of the factorization in the model case is presented in Chapter 1. In the rest of this publication, we engage a presentation of how this method may be applied to more complex situations.

In Chapter 1, we present our model problem and the formal derivation of its factorization. We refer to this derivation as formal because the validity of differentiating with respect to the parameter s of the invariant embedding has not yet been justified. We present two factorizations: one using the Dirichlet-to-Neumann operator on the mobile boundary, and the other using the Neumann-to-Dirichlet operator corresponding to a complementary domain for a boundary moving in the opposite direction. Both of these operators satisfy Riccati equations, which have the same relationship as the Riccati equations used for control and filtering. This chapter also gives an invariant embedding for the model case starting from the interior section of the cylinder: this is an example of a situation in which Ω_0 is not a subset of the boundary of Ω.

In Chapter 2, we present justifications for the computations in Chapter 1 using the Galerkin method: the function space defined on the section of the cylinder is approximated by subspaces of finite dimension, formed by the eigenfunctions of the Laplacian on this section. This is the method used by J.-L. Lions in [LIO 68a] to justify how the Riccati control equation for parabolic systems is obtained. These results cannot be applied directly to the model problem because, unlike the adjoint state operator for the control, the Dirichlet-to-Neumann operator is not bounded, giving rise to the issue of defining the meaning of the term P^2 in the Riccati equation. This method of justifying the computations of invariant embedding can probably be adapted to the diverse situations that we will study in the rest of this book. However, since writing out the full details of these adaptations would be very repetitive and long (technical developments would be needed to adapt to each situation), we will only give formal derivations in the following chapters with the exception of Chapter 4, in which we show how links to optimal control problems for parabolic systems provide a new justification for invariant embedding.

In Chapter 3, generalizations are made with respect to the boundary conditions used in the model problem. For this purpose, we present a new way of carrying out invariant embedding based on the variational formulation of the boundary value problem. We establish the relationship between the operator used for factorizing and the Steklov–Poincaré operator. The connection between the Green's functions and Hadamard's formula for a variation of the domain is also established.

In Chapter 4, we show that the model problem can be viewed as an optimal control problem of an evolution problem whose "time" variable is the abscissa along the axis of the cylinder. The state equation is a simple integrator. We recall the results obtained by J.-L. Lions in [LIO 68a] for the control of parabolic systems. A parabolic regularization of the model problem of the state equation allows us to directly use these results. We then study the convergence of regularized problem towards the original problem.

In Chapter 5, we will take another look at the method of invariant embedding applied to the factorization of boundary value problems. We will consider a discretization by finite differences of the model problem, adapted to the cylindrical geometry of the domain. The discrete analog of the invariant embedding method is then applied to the discrete problem. The result obtained, a discrete factorized problem, is exactly the same as that given by a block Gaussian LU (or Cholesky) factorization of the matrix of the discretized problem. This result also holds for other methods of discretization. We can therefore consider the factorization by invariant embedding of the boundary value problem as a "continuous" Gaussian factorization: analogous to Gaussian factorization, which transforms the initial linear system into two triangular systems (upper and lower) to be solved successively "upwards" and "downwards", the suggested method transforms the boundary value problem into two Cauchy problems to be solved successively in opposite directions. In addition, other discretizations of the factorized system give rise to other numerical schemes. We demonstrate in the Appendix that the Gaussian factorization can be obtained by the invariant embedding method. This result shows how powerful the invariant embedding method really is and sheds new light on the whole body of our work. This approach also provides a very natural explanation for the well-known condition, determining whether Gaussian factorization without pivoting can be brought to completion, namely the requirement that every principal submatrix must be non-singular. Indeed, this is a necessary condition for invariant embedding.

In Chapters 6 and 7, we aim to show that the factorization method is not limited to the Laplace operator or to cylindrical domains as in the model problem. We will begin Chapter 6 by studying a boundary value problem with a general second-order elliptic operator. In this case, additional linear terms appear in the Riccati equation for the Dirichlet-to-Neumann operator. We will then study a system of two coupled elliptic equations. Depending on the way the system is embedded, several decouplings can be obtained: if the two unknowns are embedded at once, we obtain a Riccati equation on the operator matrix connecting the two Dirichlet conditions with the two Neumann conditions; if the unknowns are successively embedded one after the other, then the equation for decoupling is more complex and in particular involves kernels coupling the two sections. In the latter case, the method allows one of the unknowns to be exactly eliminated. Here again, we can draw a connection with the Gaussian method. We will then apply the method to the linear elasticity equations. The special role played by the space variable that guides the embedding requires prior transformation of these equations before the implementation of invariant embedding. For the bilaplacian, different combinations of boundary conditions lead to several different situations, some of which are trivial factorizations, and others unstable factorizations. For the Stokes equation, the null-divergence constraint causes additional complications. Finally, we present a first approach to the way space invariant embedding could be used for an evolution problem considered on a fixed time interval.

The cylindrical domain was a relatively natural choice for our model problem, but seems very restrictive when it comes to studying general elliptic boundary value problems. In fact, the invariant embedding method only requires the construction of a family of subdomains of the original domain bounded by a family of surfaces indexed by a parameter such that this family sweeps the entire initial domain as the parameter ranges over its domain of definition. When defined in this way, invariant embedding can lead to singular situations, even for very regular surfaces, such as in the example of an initial domain that has been reduced to a point. In Chapter 7, we will first consider domains that can be reduced to a cylindrical domain by an orthogonal change of coordinates. Even though the formulas become more complicated, the computation remains similar to that for the model problem. We will then study domains that we shall call quasi-cylinders, composed of a family of plane sections whose boundaries vary continuously. With this geometry, we take on a new situation in which the field of derivation (here, naturally orthogonal to the plane sections), is not tangent to the lateral surface of the quasi-cylinder. This results in new boundary terms in the Riccati equation, which we will only derive in a formal sense. We finish this chapter with the study of a domain that is bounded by an interior and an exterior boundary of regular but arbitrary shape; invariant embedding then is carried out using a family of surfaces moving continuously from one boundary to another. These surfaces have no boundaries of their own. The proposed computation is intrinsic in the sense that it does not refer to a system of coordinates. This can give rise to a method for discretization: if we consider a finite number of surfaces discretized by finite elements, we can form a grid that is well adapted to the discretization of factorized equations obtained by this method. The computation simplifies if there is a constant number of nodes on each surface and a one-to-one correspondence between the nodes of neighboring surfaces. We will also mention the "computing zoom" method in which a family of surfaces is defined by a homothety and which allows us, from a numerical point of view, to concentrate the computation within a sub-region of interest in the domain.

In Chapter 8, we present the decomposition of the model problem into the product of an orthogonal problem and a second-order Cauchy problem.

Chapter 9 is adapted from I. Champagne's thesis [CHA 04], which uses the results established by M. Sorine. We show how the Riccati equation for the factorization of the model problem can, by means of a homographic transformation, be reduced to a linear equation. This gives a formula for representing the general solution to the Riccati equation once a particular solution is known, for example, the stationary solution. This point of view is extensively developed in finite dimensions in the book [ABO 03]. Furthermore, singularities of the homographic transformation allow us to explain and resolve difficulties previously encountered when defining the initial conditions of the Riccati equation for some boundary conditions: for example, when defining the initial condition of the equation that yields the Dirichlet-to-Neumann operator if the condition on the corresponding boundary is of Dirichlet type.

This book presents ideas previously formulated by R. Bellman and E. Angel, which are revisited and further explored within the mathematical framework developed by J.-L. Lions for partial differential equations and optimal control. The contents have already partially been published in papers. The presentation is purposely based on a very simple example in order to remain didactic. The prerequisites for this book are basic courses in functional analysis, Hilbert's theory of partial differential equations taught in third-year bachelor's or first-year master's programs in mathematics (see, for example, [BRE 99]).

One of the goals of this book is to show that the invariant embedding method used in optimal control for linear dynamic systems, which yields an optimal feedback by solving the Riccati equation, Gaussian matrix factorization and the boundary value problem factorization presented here all arise from the same methodology. This finding provides a procedure for calculating Dirichlet-to-Neumann-type operators by solving a Riccati equation. This outcome could be of use for the domain decomposition method [QUA 99] and for some inverse problems [BEN 11]. For nonlinear systems, finding the optimal feedback gives Hamilton–Jacobi–Bellman equations. As a natural continuation of our research we will transpose these ideas for factorizing nonlinear boundary value problems (for a first approach, see [JAI 75]).

We have had the privilege to exchange ideas with numerous colleagues. We would like to thank: J.-P. Yvon who co-authored the first study on the subject; M. Sorine and V. Barbu, who let us benefit of their experience with the Riccati control equation and M. Delfour for his work on the differentiation with respect to a variation in a boundary of the domain; N. Zemzemi for his collaboration on the inverse problem in electrocardiography; A. Ben Abda and B. Louro, who directed and co-directed theses on this subject; doctoral and post-doctoral students whose work on this method enabled us to extend our knowledge of the subject: I. Champagne, M. Soares, N. Belaib, M. d'Orey, F. Jday, K. Sharma, J. Bouyssier and M. Addouche.

Jacques HENRY
Angel Manuel RAMOS
August 2016

1

Presentation of the Formal Computation of Factorization

In this chapter, we introduce a "model problem", denoted by $(\mathcal{P}_0)$, of an elliptic boundary value problem, which we will use to describe the use of spatial invariant embedding and the factorized forms that follow from it. The operator for this problem is naturally the Laplacian and a cylindrical domain is assumed. We will see in Chapter 4 that this geometry arises naturally when we make an analogy with the optimal control problem for parabolic systems. However, from the point of view of solving elliptic problems, this geometry is a special case, and the generalization of the spatial invariant embedding method to other shapes of domain will be studied in Chapter 7. After defining our model problem and its functional framework, we will show how spatial invariant embedding leads to a system of decoupled evolution equations equivalent to the boundary value problem. Depending on how we choose to determine the family of problems for the invariant embedding, multiple factorizations are possible. We shall present three of them in sections 1.2, 1.3 and 1.4. In this chapter, the computation of derivatives is performed in a formal sense only. Various justifications are possible: in Chapter 2, we will use a Galerkin method and, in Chapter 4, we will consider a regularization.

1.1. Definition of the model problem and its functional framework

We will consider the Poisson problem in an open cylinder $\Omega =]0, a[\times \mathcal{O}$ of $\mathbb{R}^n$. The section $\mathcal{O}$ is a bounded open set in $\mathbb{R}^{n-1}$ with regular boundary. The axis of the cylinder is parallel to the x_1-axis. The cylinder is bounded by its faces $\Gamma_0 = \{0\} \times \mathcal{O}$, $\Gamma_a = \{a\} \times \mathcal{O}$. Here, x_1 will play a special role, comparable to that of time for the control of evolution problems. We will therefore adopt the traditional notation of evolution problems. We denote the lateral boundary of the cylinder by $\Sigma = \partial\mathcal{O}\times]0,a[$. We denote $\Delta = \sum_{i=1}^{n} \frac{\partial^2}{\partial x_i^2} = \frac{\partial^2}{\partial x_1^2} + \Delta_y$ the Laplacian, separating the variables $x_2, ..., x_n$, which we denote as y, and the variable x_1, which we will henceforth denote as x. To emphasize the special role of the x-coordinate, the function $u(x, y)$ will be written as $u(x)$ and its derivative w.r.t. x as $\dfrac{\partial u}{\partial x}$ or $\dfrac{\mathrm{d}u}{\mathrm{d}x}$. We will consider the problem:

$$(\mathcal{P}_0)\begin{cases} -\Delta u = -\dfrac{\partial^2 u}{\partial x^2} - \Delta_y u = f \quad \text{in } \Omega, \\ u|_\Sigma = 0, \quad -\dfrac{\partial u}{\partial x}|_{\Gamma_0} = u_0, \quad u|_{\Gamma_a} = u_1. \end{cases} \qquad [1.1]$$

Recall that the Sobolev space $H^{1/2}_{0,0}(\mathcal{O})$ is defined by theorem 11.7 [LIO 68b, p. 72] as the interpolation space of order $1/2$ between $H^1_0(\mathcal{O})$ and $L^2(\mathcal{O})$.

DEFINITION 1.1.– We define the spaces X, X_0 and Y as follows:

$$\begin{aligned} X &= L^2(0,a;H^1_0(\mathcal{O})) \cap H^1(0,a;L^2(\mathcal{O})), \\ X_0 &= X \cap \{u \in X : \; u|_{\Gamma_a} = 0\}, \\ Y &= \{u \in X : \frac{\partial^2 u}{\partial x^2} \in L^2(0,a;H^{-1}(\mathcal{O}))\}, \\ \hat{X} &= L^2(0,a;H^1_0(\mathcal{O})), \\ \hat{X}' &= L^2(0,a;H^{-1}(\mathcal{O})). \end{aligned} \qquad [1.2]$$

These spaces are well adapted to evolution problems. They place emphasis on the special role of the variable x. The spaces X and X_0 are respectively equipped with the norms:

$$\|u\|^2_X = \int_0^a \|u(x)\|^2_{H^1_0(\mathcal{O})} \, \mathrm{d}x + \int_0^a \|u(x)\|^2_{L^2(\mathcal{O})} \, \mathrm{d}x + \int_0^a \left\|\frac{\partial u}{\partial x}(x)\right\|^2_{L^2(\mathcal{O})} \mathrm{d}x$$

$$\|u\|^2_{X_0} = \int_0^a \|u(x)\|^2_{H^1_0(\mathcal{O})} \, \mathrm{d}x + \int_0^a \left\|\frac{\partial u}{\partial x}(x)\right\|^2_{L^2(\mathcal{O})} \mathrm{d}x.$$

The norm of X is equivalent to the norm of $H^1(\Omega)$. By the Poincaré inequality, this also applies for X_0. Consequently X_0 is closed in X, which is closed in $H^1(\Omega)$. For problem $(\mathcal{P}_0)$, we choose $u_0 \in H^{1/2}_{0,0}(\mathcal{O})'$ (the dual space of $H^{1/2}_{0,0}(\mathcal{O})$) and $u_1 \in H^{1/2}_{0,0}(\mathcal{O})$. By the Lax–Milgram theorem, it is easy to see that $(\mathcal{P}_0)$ is well-posed if f is given in X', the dual of X. Regardless, in our study, we will consider only cases where the right-hand side is defined in a smaller space $\hat{X}'$. On the one hand, this is adapted to the view of $(\mathcal{P}_0)$ as a system of coupled evolution equations:

$$-\frac{\partial u}{\partial x} = p, \quad u|_{\Gamma_1} = u_1, \qquad [1.3]$$

$$\frac{\partial p}{\partial x} - \Delta_y u = f, \quad p|_{\Gamma_0} = u_0, \qquad [1.4]$$

and, on the other hand, the regularity of f helps us overcome creating discontinuities in the invariant embedding method (remark 1.1). We will need the following result regarding traces.

PROPOSITION 1.1.– Let $u \in X$ and $\Gamma_s = \{s\} \times \mathcal{O}$, then for all $s \in [0,a]$, $s \mapsto u|_{\Gamma_s}$ belongs to $\mathcal{C}([0,a]; H^{1/2}_{0,0}(\mathcal{O}))$ and the trace mapping $u \mapsto u|_{\Gamma_s}$ is continuous and surjective from X to $H^{1/2}_{0,0}(\mathcal{O})$. If we also have $u \in Y$, then:

$$s \mapsto (u|_{\Gamma_s}, \frac{\partial u}{\partial x}|_{\Gamma_s}) \text{ belongs to } \mathcal{C}([0,a]; H^{1/2}_{0,0}(\mathcal{O}) \times H^{1/2}_{0,0}(\mathcal{O})'),$$

and the trace mapping $u \longmapsto (u|_{\Gamma_s}, \frac{\partial u}{\partial x}|_{\Gamma_s})$ is continuous and surjective from Y to $H^{1/2}_{0,0}(\mathcal{O}) \times H^{1/2}_{0,0}(\mathcal{O})'$.

PROOF.– This is a direct application of theorem 3.1 from [LIO 68b, p.23], taking into account that $H^{1/2}_{0,0}(\mathcal{O})'$ is the interpolation space of order $1/2$ between $L^2(\mathcal{O})$ and $H^{-1}(\mathcal{O})$ (see theorem 12.4 in [LIO 68b, p. 81]). □

The variational formulation of problem $(\mathcal{P}_0)$ is:

$$a(u,v) = \int_\Omega \nabla u \nabla v \, dx \, dy =$$
$$=< f, v >_{\hat{X}' \times \hat{X}} + < u_0, v|_{\Gamma_0} >_{H^{1/2}_{0,0}(\mathcal{O})' \times H^{1/2}_{0,0}(\mathcal{O})}, \quad \forall v \in X_0, \qquad [1.5]$$

where $u \in X$ satisfies the constraint $u|_{\Gamma_a} = u_1$. After translation of a lifting of u_1, we can apply the Lax–Milgram theorem. We find that $(\mathcal{P}_0)$ is well-posed and has a solution in X.

1.2. Direct invariant embedding

We will use the invariant embedding method introduced by R. Bellman (see [BEL 57a]). In control theory, this method is used to deduce the optimal feedback: the control problem is embedded in a family of identical problems defined between the current time and the final time, where the initial state is the state observed at the present moment. Here, invariant embedding is performed spatially: problem $(\mathcal{P}_0)$ is embedded in a family of similar problems $(\mathcal{P}_{s,h})$ defined on $\Omega_s =]0, s[\times \mathcal{O}$ for $s \in]0, a]$ (see Figure 1.1). For each problem, we impose the Dirichlet boundary condition $u_{s,h}|_{\Gamma_s} = h$. Each problem is dependent on both s and h:

$$(\mathcal{P}_{s,h}) \begin{cases} -\Delta u_{s,h} = f \quad \text{in } \Omega_s, \\ u_{s,h}|_\Sigma = 0, \quad -\dfrac{\partial u_{s,h}}{\partial x}|_{\Gamma_0} = u_0, \quad u_{s,h}|_{\Gamma_s} = h. \end{cases} \qquad [1.6]$$

It is clear that $(\mathcal{P}_0)$ is exactly $(\mathcal{P}_{s,h})$ for $s = a, h = u_1$, and so $u = u_{a,u_1}$.

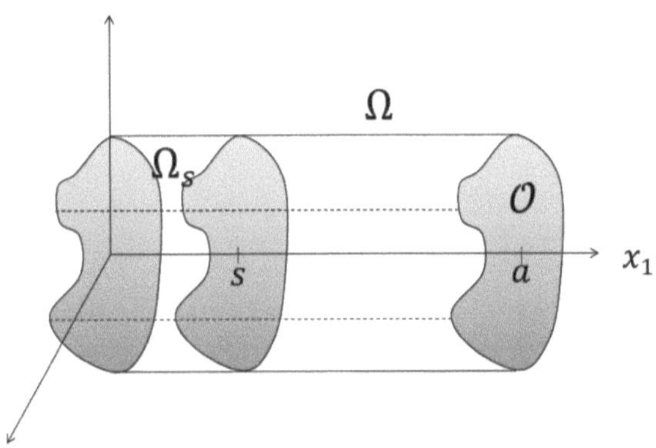

Figure 1.1. *The domain of the problem and subdomains for the direct invariant embedding*

For $s \in]0, a]$ we have defined $Y_s = \{v \in H^1(\Omega_s) : \frac{\partial^2 v}{\partial x^2} \in L^2(0, s; H^{-1}(\mathcal{O}))$ and $v|_\Sigma = 0\}$. The calculation is now similar to that of

Lions [LIO 68a] used to obtain the optimal feedback of a quadratic linear optimal control problem for a parabolic equation (which we shall recall in Chapter 4). Therefore, we separate $u_{s,h}$ into a linear part in h, $\gamma_{s,h}$, and a part that is independent of h, β_s.

DEFINITION 1.2.– For all $s \in]0,a]$, we define the Dirichlet-to-Neumann (DtN) mapping $P(s)$ by $P(s)h = \frac{\partial \gamma_{s,h}}{\partial x}|_{\Gamma_s}$, where $h \in H^{1/2}_{0,0}(\mathcal{O})$ and $\gamma_{s,h} \in Y_s$ is the solution of:

$$\begin{cases} -\Delta\gamma_{s,h} = 0 \quad \text{in } \Omega_s, \\ \gamma_{s,h}|_\Sigma = 0, \quad \dfrac{\partial \gamma_{s,h}}{\partial x}|_{\Gamma_0} = 0, \quad \gamma_{s,h}|_{\Gamma_s} = h. \end{cases} \tag{1.7}$$

We set $P(0)h = 0$ and define w by $w(s) = \frac{\partial \beta_s}{\partial x}|_{\Gamma_s}$, where $\beta_s \in Y_s$ is the solution of:

$$\begin{cases} -\Delta\beta_s = f \quad \text{in } \Omega_s, \\ \beta_s|_\Sigma = 0, \quad -\dfrac{\partial \beta_s}{\partial x}|_{\Gamma_0} = u_0, \quad \beta_s|_{\Gamma_s} = 0. \end{cases} \tag{1.8}$$

We also set $w(0) = -u_0$.

To simplify notation and whenever the reference to the affine or linear dependency on h is not necessary, we write $u_s = u_{s,h}$ and $\gamma_s = \gamma_{s,h}$. For all $s \in [0,a]$, $P(s) : H^{1/2}_{0,0}(\mathcal{O}) \to H^{1/2}_{0,0}(\mathcal{O})'$ is a linear operator and $w(s) = \dfrac{\mathrm{d}\beta_s(s)}{\mathrm{d}x} \in H^{1/2}_{0,0}(\mathcal{O})'$ because the problems on β_s and γ_s are well-posed, as we mentioned at the end of section 1.1 in proposition 1.1.

PROPOSITION 1.2.– The linear operator $P(s) : H^{1/2}_{0,0}(\mathcal{O}) \mapsto H^{1/2}_{0,0}(\mathcal{O})'$, is continuous and self-adjoint for all $s \in [0,a]$. The associated bilinear form on $H^{1/2}_{0,0}(\mathcal{O})$, $< P(s)h, \overline{h} >_{H^{1/2}_{0,0}(\mathcal{O})' \times H^{1/2}_{0,0}(\mathcal{O})}$, is coercive for all $s \in]0,a]$. Furthermore, there exists $C > 0$ independent of a such that:

$$\frac{s}{C+s}$$

is a constant of coercivity.

PROOF.– The operator $P(s)$ is defined as the composition of two continuous operators: $h \mapsto \gamma_s \mapsto \dfrac{\partial \gamma_s}{\partial x}|_{\Gamma_s}$ of $\mathcal{L}(H^{1/2}_{0,0}(\mathcal{O}), Y_s)$ and $\mathcal{L}(Y_s, H^{1/2}_{0,0}(\mathcal{O})')$, respectively, and is therefore continuous. Let $h, \overline{h} \in H^{1/2}_{0,0}(\mathcal{O})$ and $\gamma_s, \overline{\gamma}_s$ be the corresponding solutions in Ω_s. By Green's formula, we have that:

$$0 = -\int_{\Omega_s} \Delta\gamma_s \overline{\gamma}_s \, \mathrm{d}x \, \mathrm{d}y = $$
$$= \int_{\Omega_s} \nabla\gamma_s \nabla\overline{\gamma}_s \, \mathrm{d}x \, \mathrm{d}y - < P(s)h, \overline{h} >_{H^{1/2}_{0,0}(\mathcal{O})' \times H^{1/2}_{0,0}(\mathcal{O})}, \qquad [1.9]$$

which shows self-adjointness of $P(s)$ by permuting h with $\overline{h}$. Setting $h = \overline{h}$ we have:

$$< P(s)h, h >_{H^{1/2}_{0,0}(\mathcal{O})' \times H^{1/2}_{0,0}(\mathcal{O})} = \int_{\Omega_s} \left(|\nabla_y \gamma_s|^2 + \left| \frac{\partial \gamma_s}{\partial x} \right|^2 \right) \mathrm{d}x \; \mathrm{d}y. \qquad [1.10]$$

Therefore, by a calculation similar to the one used to prove the Poincaré inequality:

$$\|h\|^2_{H^{1/2}_{0,0}(\mathcal{O})} = \|\gamma_s(t)\|^2_{H^{1/2}_{0,0}(\mathcal{O})} + \int_t^s \frac{\mathrm{d}}{\mathrm{d}x} \|\gamma_s(x)\|^2_{H^{1/2}_{0,0}(\mathcal{O})} \, \mathrm{d}x, \qquad [1.11]$$

and

$$\frac{\mathrm{d}}{\mathrm{d}x} \|\gamma_s(x)\|^2_{H^{1/2}_{0,0}(\mathcal{O})} = 2 < \frac{\mathrm{d}\gamma_s(x)}{\mathrm{d}x}, \gamma_s(x) >_{L^2(\mathcal{O}) \times H^1_0(\mathcal{O})}, \qquad [1.12]$$

where $<,>_{L^2(\mathcal{O}) \times H^1_0(\mathcal{O})}$ is the duality pairing with $H^{1/2}_{0,0}(\mathcal{O})$ as the pivot space. Then, taking into account the continuous injection of $H^{1/2}_{0,0}(\mathcal{O})$ in $H^1_0(\mathcal{O})$, there exists $C > 0$ that is dependent of $\mathcal{O}$ but not of a such that:

$$\|h\|^2_{H^{1/2}_{0,0}(\mathcal{O})} \leq C \|\gamma_s(t)\|^2_{H^1_0(\mathcal{O})} + \int_t^s \|\gamma_s(x)\|^2_{H^1_0(\mathcal{O})} \, \mathrm{d}x +$$
$$+ \int_t^s \left\| \frac{\mathrm{d}\gamma_s(x)}{\mathrm{d}x} \right\|^2_{L^2(\mathcal{O})} \mathrm{d}x \qquad [1.13]$$

for $0 \leq t \leq s \leq a$. Integrating with respect to t from 0 to s, we obtain:

$$s \|h\|^2_{H^{1/2}_{0,0}(\mathcal{O})} \leq (C + s) \int_0^s \|\gamma_s(x)\|^2_{H^1_0(\mathcal{O})} \, \mathrm{d}x + s \int_0^s \left\| \frac{\mathrm{d}\gamma_s(x)}{\mathrm{d}x} \right\|^2_{L^2(\mathcal{O})} \mathrm{d}x \qquad [1.14]$$

and consequently:

$$< P(s)h, h >_{H^{1/2}_{0,0}(\mathcal{O})' \times H^{1/2}_{0,0}(\mathcal{O})} \geq \frac{s}{C+s} \parallel h \parallel^2_{H^{1/2}_{0,0}(\mathcal{O})}, \qquad [1.15]$$

which proves coercivity. □

From the linearity of $(\mathcal{P}_{s,h})$, we have:

$$\frac{\partial u_{s,h}}{\partial x}|_{\Gamma_s} = P(s)h + w(s), \qquad [1.16]$$

where $u_{s,h} = \beta_s + \gamma_{s,h}$, and the subscript s, h reminds us that the problem is defined on Ω_s and it depends on the data h.

We will now consider the restriction of $u_{s,h}$ to Ω_x for $x \leq s$. This is the solution of problem $(\mathcal{P}_{x,u_{s,h}(x)})$. From the reasoning given above, we have the relation:

$$\frac{\mathrm{d}u_{s,h}}{\mathrm{d}x}(x) = P(x)u_{s,h}(x) + w(x). \qquad [1.17]$$

Continuing the analogy with evolution problems, we call trajectory the mapping:

$$x \longmapsto u_{s,h}(x) \in H^{1/2}_{0,0}(\mathcal{O}).$$

Therefore, [1.17] is an identity for the family of trajectories that depend on h. Each trajectory is continuous from $[0, s]$ with values in $H^{1/2}_{0,0}(\mathcal{O})$ for $0 < s \leq a$, and is also defined in $L^2(0, s; H^1_0(\mathcal{O}))$. We can therefore, at least formally, take the derivative of [1.17] with respect to x, and using the equation satisfied by u_s:

$$\frac{\partial^2 u_s}{\partial x^2} = -\Delta_y u_s - f = \frac{\mathrm{d}P}{\mathrm{d}x} u_s + P\frac{\mathrm{d}u_s}{\mathrm{d}x} + \frac{\mathrm{d}w}{\mathrm{d}x}. \qquad [1.18]$$

Substituting $\frac{\mathrm{d}u_s}{\mathrm{d}x}$ from [1.17]:

$$0 = \left(\frac{\mathrm{d}P(x)}{\mathrm{d}x} + P^2(x) + \Delta_y\right) u_s(x)$$
$$+\frac{\mathrm{d}w(x)}{\mathrm{d}x} + P(x)w(x) + f(x) \text{ for } 0 \leq x \leq s.$$

Setting $x = s$:

$$0 = \left(\frac{\mathrm{d}P}{\mathrm{d}x}(s) + P^2(s) + \Delta_y\right) h + \frac{\mathrm{d}w}{\mathrm{d}x}(s) + P(s)w(s) + f(s).$$

This computation applies for all trajectories that satisfy the Poisson equation and the boundary condition in Γ_0, $-\frac{\partial u}{\partial x}|_{\Gamma_0} = u_0$ (this trajectory is entirely determined if we know h and s). Therefore, h and s are arbitrary and, from the previous identity, we deduce:

$$\frac{\mathrm{d}P}{\mathrm{d}x} + P^2 = -\Delta_y, \text{ for } x \in]0, a], \qquad [1.19]$$

and for the residual term:

$$\frac{\mathrm{d}w}{\mathrm{d}x} + Pw = -f, \text{for } x \in]0, a]. \qquad [1.20]$$

REMARK 1.1.– Referring to the classical theory of parabolic evolution equations in weak form, we can see the benefit of choosing the right-hand side f in $L^2(0, a : \hat{X}')$. In particular, it avoids discontinuities in w and consequently in $\frac{\partial u}{\partial x}$.

In Chapter 2, we will justify that this argument of an arbitrary trajectory can be extended by continuity up to $x = 0$. Hence, evaluating relation [1.17] at $x = 0$ and using the fact that $u(0)$ is arbitrary gives the initial conditions for P and w:

$$P(0) = 0, \quad w(0) = -u_0. \qquad [1.21]$$

We wish to emphasize that P is an operator dependent on x acting on functions of the variable y defined on the section $\mathcal{O}$. It satisfies the Riccati equation [1.19] , which has the property of not having any linear terms. This property is specific to the Laplacian in a cylinder and does not hold in other cases (see examples in Chapter 6). The right-hand side is an unbounded operator on $L^2(\mathcal{O})$. Combining equations [1.19–1.20], the initial conditions for P and w [1.21], equation [1.17] and the condition on Γ_a for u, we obtain the system:

$$\frac{\mathrm{d}P}{\mathrm{d}x} + P^2 + \Delta_y = 0, \qquad P(0) = 0, \qquad [1.22]$$

$$\frac{\mathrm{d}w}{\mathrm{d}x} + Pw = -f, \qquad w(0) = -u_0, \qquad [1.23]$$

$$-\frac{\mathrm{d}u}{\mathrm{d}x} + Pu = -w, \qquad u(a) = u_1. \qquad [1.24]$$

The system of equations given by [1.22–1.24] is decoupled because we can integrate the first two equations with respect to x from 0 to a to get P and w ("*downward phase*"), then u is obtained by backward integration of the third equation ("*upward phase*"). Formally, knowing P, we have factorized $(\mathcal{P}_0)$ into:

$$-\left(\frac{\mathrm{d}}{\mathrm{d}x} + P\right)\left(\frac{\mathrm{d}}{\mathrm{d}x} - P\right) u = f. \qquad [1.25]$$

In [1.22], by abuse of notation, we wrote $-\Delta_y$ for the abstract operator of the boundary value problem relative to the Laplacian with Dirichlet boundary conditions, which is an isomorphism from $H_0^1(\mathcal{O})$ to $H^{-1}(\mathcal{O})$.

We shall show in Chapter 2 the following theorem that gives a precise meaning to [1.22–1.24].

THEOREM 1.1.– The solution u of $(\mathcal{P}_0)$ is the unique solution of system [1.22–1.24] of decoupled Cauchy problems of first order in x, in the following sense:

1) the operator P, defined in definition 1.2 is self-adjoint, non-negative and has the following properties:

$$P \in L^\infty\big(0, a; \mathcal{L}\big(H_0^1(\mathcal{O}), L^2(\mathcal{O})\big) \cap \mathcal{L}(H_{0,0}^{1/2}(\mathcal{O}), H_{0,0}^{1/2}(\mathcal{O})') \\ \cap \mathcal{L}(L^2(\mathcal{O}), H^{-1}(\mathcal{O}))\big).$$

It satisfies the following Riccati equation in the weak sense:

$$\left(\frac{\mathrm{d}P}{\mathrm{d}x}h, \bar{h}\right) + (Ph, P\bar{h}) = (\nabla_y h, \nabla_y \bar{h}) \quad \forall\, h, \bar{h} \in H_0^1(\mathcal{O}), \qquad [1.26]$$

almost everywhere in $]0, a[$, with the initial condition $P(0) = 0$;

2) the function w defined in definition 1.2 satisfies the equation:

$$\left\langle \frac{\mathrm{d}w}{\mathrm{d}x}, h \right\rangle + \langle Pw, h\rangle = -\langle f, h\rangle \quad \forall\, h \in H_0^1(\mathcal{O}), \qquad [1.27]$$

almost everywhere in $]0, a[$, with the boundary condition $w(0) = -u_0$, where $\langle .,. \rangle$ is the duality pairing between $H_0^1(\mathcal{O})$ and $H^{-1}(\mathcal{O})$. Defined this way, the problem is well-posed for w in $L^2(0, a; L^2(\mathcal{O}))$ and $\frac{\mathrm{d}w}{\mathrm{d}x} \in L^2(0, a; H^{-1}(\mathcal{O}))$, and therefore $w \in \mathcal{C}([0, a]; H_{0,0}^{1/2}(\mathcal{O})')$;

3) the function $u \in X$ satisfies the equation:

$$-\left(\frac{\mathrm{d}u}{\mathrm{d}x}, h\right) + (Pu, h) = -(w, h) \quad \forall\, h \in L^2(\mathcal{O}), \qquad [1.28]$$

almost everywhere in $]0, a[$, with the initial condition $u(a) = u_1$. The problem thus defined is well-posed in the direction of decreasing x.

Even though P is non-local, the equations in w and u are of parabolic type with respect to x and the corresponding Cauchy problem is well-posed. Furthermore, the factorization [1.25] is self-adjoint because each term is adjoint to the other.

We will show in Chapter 5 that a discretized version of this computation corresponds exactly to a block Gaussian LU factorization. Reciprocally, we can say that this invariant embedding computation is in fact a continuous Gaussian factorization of the boundary value problem.

REMARK 1.2.– We can verify the factorization [1.25] by evaluating the product of the two terms and using the Riccati equation [1.19]. Regardless, it should be noted that the factorized problem continues to satisfy the boundary conditions. Therefore, we have factorized the boundary value problem and this factorization is different from the factorization of the partial derivative operator:

$$-\Delta = -\left(\frac{\mathrm{d}}{\mathrm{d}x} + (-\Delta_y)^{1/2}\right)\left(\frac{\mathrm{d}}{\mathrm{d}x} - (-\Delta_y)^{1/2}\right).$$

In this calculation, the boundary conditions in $x = 0$ and $x = a$ are not taken into account.

REMARK 1.3.– We can also verify from the form [1.25] that the boundary conditions are satisfied by u. The condition on Σ is reflected in the fact that u belongs to $H^{1/2}_{0,0}(\mathcal{O})$ for almost all x. The condition on Γ_a is the initial condition for integration of the third equation [1.25]. We formally recover the condition on Γ_0 by considering the third equation [1.25] for $x = 0$ and using the initial conditions $P(0) = 0$ and $w(0) = -u_0$.

REMARK 1.4.– We also make the remark that in [1.22], [1.23] and [1.24], the Riccati equation on P depends only on the partial differential operator of $(\mathcal{P}_0)$. Furthermore, once P is known, the equations on w and u depend only on the data: boundary conditions and right-hand side. Therefore, if we have to solve the same problem many times for different values, the factorized form [1.25] reduces the problem to the integration of just two decoupled parabolic Cauchy problems for each new set of data. If only u_1 is modified, only the equation on u needs to be integrated again.

REMARK 1.5.– In finite dimensions, we know how to explicitly calculate the solution of the Riccati equation [1.22]. Therefore, we can non-rigorously write:

$$P(x) = (-\Delta_y)^{\frac{1}{2}}\left(\exp(2x(-\Delta_y)^{\frac{1}{2}}) - I\right)\left(\exp(2x(-\Delta_y)^{\frac{1}{2}}) + I\right)^{-1}. \quad [1.29]$$

This formula was used in [BOU 08] to justify the previous computations using a Yosida regularization.

REMARK 1.6.– As we stated in the beginning of this section, this computation is analogous to the classical computation in automatics for obtaining the optimal control in "closed loop". Here, as soon as we know $u|_{\Gamma_s}$, we can obtain u in Ω_s and $\frac{\partial u}{\partial x}|_{\Gamma_s}$. In automatics, when the state at the present moment is not measured and the system might have gone through perturbations, we use the Kalman filter to reconstruct the present state using observations on its history. We can imagine a similar strategy for reconstructing $u|_{\Gamma_s}$ if, for example, the condition on Γ_a is not known but we have measurements of the state in a complementary part of the domain: $\widetilde{\Omega}_s =]s, a[\times \mathcal{O}$.

1.3. Backward invariant embedding

Naturally, the solution u can also be obtained by invariant embedding in a complementary part of the domain: $\widetilde{\Omega}_s =]s, a[\times \mathcal{O}$.

For $s \in [0, a[$, we define $\widetilde{Y}_s = \{v \in H^1(\widetilde{\Omega}_s) : \frac{\partial^2 v}{\partial x^2} \in L^2(s, a; H^{-1}(\mathcal{O}))$ and $v_{|\Gamma_a} = v_{|\Sigma} = 0\}$ (see Figure 1.2) and:

$$(\widetilde{\mathcal{P}}_{s,h}) \begin{cases} -\Delta \tilde{u}_{s,h} = f \quad \text{in } \widetilde{\Omega}_s, \\ \tilde{u}_{s,h}|_\Sigma = 0, \quad -\dfrac{\partial \tilde{u}_{s,h}}{\partial x}|_{\Gamma_s} = h, \quad \tilde{u}_{s,h}|_{\Gamma_a} = u_1, \end{cases}$$

where $h \in H^{1/2}_{0,0}(\mathcal{O})'$.

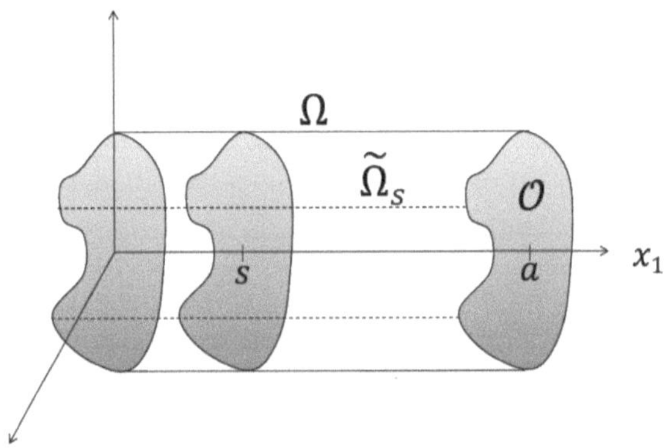

Figure 1.2. *Subdomains for backward invariant embedding*

DEFINITION 1.3.– For every $s \in [0, a[$, we define the Neumann-to-Dirichlet mapping (NtD) $Q(s) \in \mathcal{L}(H^{1/2}_{0,0}(\mathcal{O})', H^{1/2}_{0,0}(\mathcal{O}))$ by $Q(s)h = \tilde{\gamma}_{s,h}|_{\Gamma_s}$, where $h \in H^{1/2}_{0,0}(\mathcal{O})'$ and $\tilde{\gamma}_{s,h} \in \widetilde{Y}_s$ is the solutions of:

$$\begin{cases} -\Delta \tilde{\gamma}_{s,h} = 0 \quad \text{in } \widetilde{\Omega}_s, \\ \tilde{\gamma}_{s,h}|_\Sigma = 0, \quad -\dfrac{\partial \tilde{\gamma}_{s,h}}{\partial x}|_{\Gamma_s} = h, \quad \tilde{\gamma}_{s,h}|_{\Gamma_a} = 0. \end{cases} \qquad [1.30]$$

We set $Q(a)h = 0$ and define r by $r(s) = \tilde{\beta}_s|_{\Gamma_s} \in H^{1/2}_{0,0}(\mathcal{O})$, where $\tilde{\beta}_s \in \widetilde{Y}_s$ is the solution of:

$$\begin{cases} -\Delta\tilde{\beta}_s = f \quad \text{in } \widetilde{\Omega}_s, \\ \tilde{\beta}_s|_\Sigma = 0, \quad \dfrac{\partial\tilde{\beta}_s}{\partial x}|_{\Gamma_s} = 0, \quad \tilde{\beta}_s|_{\Gamma_a} = u_1. \end{cases} \qquad [1.31]$$

We also set $r(a) = u_1$.

We obtain:

$$\tilde{u}_{s,h}(s) = Q(s)h + r(s). \qquad [1.32]$$

As before, when it is not necessary to stress the dependency on h, we can simplify our notation

$$\tilde{u}_s = \tilde{u}_{s,h}, \quad \tilde{\gamma}_s = \tilde{\gamma}_{s,h}.$$

Let us show that Q has similar properties to those obtained in the previous section.

PROPOSITION 1.3.– The linear operator $Q(s) : H^{1/2}_{0,0}(\mathcal{O})' \to H^{1/2}_{0,0}(\mathcal{O})$ is continuous, self-adjoint and non-negative for all $s \in [0, a]$. Its associated bilinear form on $H^{1/2}_{0,0}(\mathcal{O})'$, $< Q(s)h, \overline{h} >_{H^{1/2}_{0,0}(\mathcal{O}) \times H^{1/2}_{0,0}(\mathcal{O})'}$ is coercive for all $s \in [0, a[$.

PROOF.– The operator $Q(s)$ is defined by the composition of two continuous operators: $h \to \tilde{\gamma}_s \to \tilde{\gamma}_s|_{\Gamma_s}$ in $\mathcal{L}(H^{1/2}_{0,0}(\mathcal{O})', \widetilde{Y}_s)$ and $\mathcal{L}(\widetilde{Y}_s, H^{1/2}_{0,0}(\mathcal{O}))$, respectively, and is therefore continuous. Let $h, \overline{h} \in H^{1/2}_{0,0}(\mathcal{O})'$ and $\tilde{\gamma}_s, \overline{\tilde{\gamma}}_s$ be the corresponding solutions of [1.30] in $\widetilde{\Omega}_s$. We have, by Green's formula:

$$0 = -\int_{\widetilde{\Omega}_s} \tilde{\gamma}_s \Delta\overline{\tilde{\gamma}}_s \, dx\, dy = \int_{\widetilde{\Omega}_s} \nabla\tilde{\gamma}_s \nabla\overline{\tilde{\gamma}}_s \, dx dy - $$
$$- < Q(s)h, \overline{h} >_{H^{1/2}_{0,0}(\mathcal{O}) \times H^{1/2}_{0,0}(\mathcal{O})'}, \qquad [1.33]$$

which shows the self-adjointness of $Q(s)$ if we permute h and $\overline{h}$ and its non-negativity by setting $h = \overline{h}$. We have:

$$< Q(s)h, h >_{H_{0,0}^{1/2}(\mathcal{O})\times H_{0,0}^{1/2}(\mathcal{O})'} = \int_{\widetilde{\Omega}_s} \left(|\nabla_y \tilde{\gamma}_s|^2 + \left| \frac{\partial \tilde{\gamma}_s}{\partial x} \right|^2 \right) \mathrm{d}x \, \mathrm{d}y. \qquad [1.34]$$

Therefore, by a computation similar to the one we used in proposition 1.2

$$\begin{aligned}
\|h\|^2_{H_{0,0}^{1/2}(\mathcal{O})'} &= \left\| \frac{\mathrm{d}\tilde{\gamma}_s}{\mathrm{d}x}(t) \right\|^2_{H_{0,0}^{1/2}(\mathcal{O})'} - \int_s^t \frac{\mathrm{d}}{\mathrm{d}x} \left\| \frac{\mathrm{d}\tilde{\gamma}_s}{\mathrm{d}x}(x) \right\|^2_{H_{0,0}^{1/2}(\mathcal{O})'} \mathrm{d}x \\
&\leq C \left\| \frac{\mathrm{d}\tilde{\gamma}_s}{\mathrm{d}x}(t) \right\|^2_{L^2(\mathcal{O})} + \int_s^t \left\| \frac{\mathrm{d}\tilde{\gamma}_s}{\mathrm{d}x}(x) \right\|^2_{L^2(\mathcal{O})} \mathrm{d}x \\
&\quad + \int_s^t \left\| \frac{\mathrm{d}^2\tilde{\gamma}_s}{\mathrm{d}x^2}(x) \right\|^2_{H^{-1}(\mathcal{O})} \mathrm{d}x,
\end{aligned}$$

for $0 \leq s \leq t \leq a$. Now:

$$\left\| \frac{\mathrm{d}^2\tilde{\gamma}_s}{\mathrm{d}x^2}(x) \right\|^2_{H^{-1}(\mathcal{O})} = \|\Delta_y \tilde{\gamma}_s(x)\|^2_{H^{-1}(\mathcal{O})} \leq C \, \|\tilde{\gamma}_s(x)\|^2_{H_0^1(\mathcal{O})},$$

from which:

$$\begin{aligned}
\|h\|^2_{H_{0,0}^{1/2}(\mathcal{O})'} \leq C \left\| \frac{\mathrm{d}\tilde{\gamma}_s}{\mathrm{d}x}(t) \right\|^2_{L^2(\mathcal{O})} + \int_s^t \left\| \frac{\mathrm{d}\tilde{\gamma}_s}{\mathrm{d}x}(x) \right\|^2_{L^2(\mathcal{O})} \mathrm{d}x + \\
+ C \int_s^t \|\tilde{\gamma}_s(x)\|^2_{H_0^1(\mathcal{O})} \mathrm{d}x.
\end{aligned} \qquad [1.35]$$

Integrating with respect to t from s to a, we obtain:

$$\begin{aligned}
(a-s) \, \|h\|^2_{H_{0,0}^{1/2}(\mathcal{O})'} \leq (C + a - s) \int_s^a \left\| \frac{\mathrm{d}\tilde{\gamma}_s}{\mathrm{d}x}(x) \right\|^2_{L^2(\mathcal{O})} \mathrm{d}x + \\
+ (a-s) C \int_s^a \|\tilde{\gamma}_s(x)\|^2_{H_0^1(\mathcal{O})} \mathrm{d}x
\end{aligned} \qquad [1.36]$$

and, consequently, assuming $C > 1$ (or by replacing C with 1):

$$< Q(s)h, h >_{H^{1/2}_{0,0}(\mathcal{O}) \times H^{1/2}_{0,0}(\mathcal{O})'} \geq \frac{a-s}{C(C+a-s)} \parallel h \parallel^2_{H^{1/2}_{0,0}(\mathcal{O})'}, \quad [1.37]$$

which proves coercivity for all $s \in [0, a[$. □

We will now call trajectory the mapping:

$$x \longmapsto \frac{\mathrm{d}\tilde{u}_s}{\mathrm{d}x}(x) \in H^{1/2}_{0,0}(\mathcal{O})', \quad x \in [s, a[,$$

where $\tilde{u}_s$ satisfies $(\widetilde{\mathcal{P}}_{s,h})$. Each trajectory is determined by a pair (s, h). Along a trajectory, we have the identity:

$$\tilde{u}_s(x) = -Q(x)\frac{\mathrm{d}\tilde{u}_s}{\mathrm{d}x}(x) + r(x). \quad [1.38]$$

Formally differentiating with respect to x gives:

$$\frac{\mathrm{d}\tilde{u}_s}{\mathrm{d}x}(x) = -\frac{\mathrm{d}Q}{\mathrm{d}x}(x)\frac{\mathrm{d}\tilde{u}_s}{\mathrm{d}x}(x) - Q(x)\frac{\mathrm{d}^2\tilde{u}_s}{\mathrm{d}x^2}(x) + \frac{\mathrm{d}r}{\mathrm{d}x}(x).$$

Using $(\widetilde{\mathcal{P}}_{s,h})$ and substituting as in [1.38], we obtain:

$$\frac{\mathrm{d}\tilde{u}_s}{\mathrm{d}x}(x) = \left(-\frac{\mathrm{d}Q}{\mathrm{d}x} - Q\Delta_y Q\right)\frac{\mathrm{d}\tilde{u}_s}{\mathrm{d}x}(x) + \frac{\mathrm{d}r}{\mathrm{d}x}(x) + Q\Delta_y r(x) + Qf(x). \quad [1.39]$$

Setting $x = s$ and using the fact that $\frac{\mathrm{d}\tilde{u}_{s,h}}{\mathrm{d}x}(s) = h$ is arbitrary, we obtain the decoupled system:

$$\frac{\mathrm{d}Q}{\mathrm{d}x} + Q\Delta_y Q + I = 0, \qquad Q(a) = 0, \quad [1.40]$$

$$\frac{\mathrm{d}r}{\mathrm{d}x} + Q\Delta_y r = -Qf, \qquad r(a) = u_1, \quad [1.41]$$

$$Q\frac{\mathrm{d}u}{\mathrm{d}x} + u = r, \qquad u(0) = Q(0)u_0 + r(0). \quad [1.42]$$

The variational formulation corresponding to [1.26] for P is:

$$\left(\frac{\mathrm{d}Q}{\mathrm{d}x}u, v\right) - (\nabla_y Qu, \nabla_y Qv) = -(u, v),$$
$$\forall u, v \in L^2(\mathcal{O}), \quad Q(a) = 0. \qquad [1.43]$$

The initial condition for u is obtained from [1.38] for $x = 0$. We have formally obtained the factorization of $(\mathcal{P}_0)$:

$$-\left(\frac{\mathrm{d}}{\mathrm{d}x} + Q\Delta_y\right) \circ \left(I + Q\frac{\mathrm{d}}{\mathrm{d}x}\right) u = Qf. \qquad [1.44]$$

To obtain a self-adjoint form of this factorization, we use the invertibility of Q for $x \neq a$ and its self-adjointness, which was shown in proposition 1.3. The composition of the left-hand side of [1.44] with Q^{-1} yields:

$$-Q^{-1}\circ\left(\frac{\mathrm{d}}{\mathrm{d}x} + Q\circ\Delta_y\right) \circ \left(I + Q\circ\frac{\mathrm{d}}{\mathrm{d}x}\right) =$$
$$= -\,Q^{-1}\circ\left(\frac{\mathrm{d}}{\mathrm{d}x} + Q\circ\Delta_y\right)\circ Q \circ\left(Q^{-1} + \frac{\mathrm{d}}{\mathrm{d}x}\right)$$
$$= -\left(Q^{-1}\circ\frac{\mathrm{d}}{\mathrm{d}x}\circ Q + \Delta_y\circ Q\right)\circ\left(Q^{-1} + \frac{\mathrm{d}}{\mathrm{d}x}\right)$$
$$= \left(-\frac{\mathrm{d}}{\mathrm{d}x} + Q^{-1}\right)\circ\left(Q^{-1} + \frac{\mathrm{d}}{\mathrm{d}x}\right), \qquad [1.45]$$

because of the Riccati equation on Q [1.40] (see also remark 1.7).

REMARK 1.7.– In this computation, we used the symbol $\circ$ for the composition of operators. This notation is convenient because it avoids confusion when the first operator is the derivative with respect to x, in particular $\frac{\mathrm{d}}{\mathrm{d}x}\circ Q(x) \neq \frac{\mathrm{d}Q(x)}{\mathrm{d}x}$. Indeed, we have:

$$\frac{\mathrm{d}}{\mathrm{d}x}\circ Q(x) = \frac{\mathrm{d}Q(x)}{\mathrm{d}x} + Q(x)\circ\frac{\mathrm{d}}{\mathrm{d}x}. \qquad [1.46]$$

From now on, we will use this notation only in situations like this, otherwise we will omit the symbol to lighten notation.

REMARK 1.8.– The operator Q^{-1} has a singularity at $x = a$. We will come back to this point in section 3.2.2 and in Chapter 9, as well as from a numerical perspective in remark 5.2.

REMARK 1.9.– In Chapter 4, we will show that the Riccati equation [1.22] may be interpreted as the result of decoupling an optimal control problem. Hence, the relation between the Riccati equations [1.22] and [1.40] is the same as the one between the Riccati equations for control and filtering: the derivative changes sign and the operators I and $-\Delta_y$ are interchanged.

REMARK 1.10.– For the problem studied here, with the Laplace operator and a cylindrical domain, we can substitute the variable x with $a - x$ without changing problem $(\mathcal{P}_0)$, except for the boundary conditions on Γ_0 and Γ_a. Actually, these boundary conditions are interchanged. We see that there exists a relation between $P(x)$ satisfying [1.22] and $Q(a - x)$ for Q satisfying [1.40]. Indeed, we have the relation:

$$P(x) = -Q(a - x)\Delta_y. \tag{1.47}$$

To show this relation, we right-multiply the equation satisfied by $Q'(x) = Q(a - x)$:

$$\frac{\mathrm{d}Q'}{\mathrm{d}x} = Q'\Delta_y Q' + I, \quad Q'(0) = 0, \tag{1.48}$$

by $-\Delta_y$. We have:

$$-\frac{\mathrm{d}(Q'\Delta_y)}{\mathrm{d}x} = -Q'\Delta_y Q'\Delta_y - \Delta_y, \quad Q'(0)\Delta_y = 0. \tag{1.49}$$

By comparison with [1.22] and by the uniqueness of the solution of this equation, we obtain relation [1.47]. Using [1.29], we deduce the “improper” explicit formula for Q:

$$\begin{aligned} Q(x) = (-\Delta_y)^{-\frac{1}{2}} &\left(\exp(2(a - x)(-\Delta_y)^{\frac{1}{2}}) - I\right) \\ &\left(\exp(2(a - x)(-\Delta_y)^{\frac{1}{2}}) + I\right)^{-1}. \end{aligned} \tag{1.50}$$

1.4. Internal invariant embedding

Another way of looking at invariant embedding is to consider a family of "internal" sub-cylinders. They are internal in the sense that they grow from a section of the cylinder other than Γ_0 or Γ_a. This is an example of a situation where invariant embedding is based on a family of domains for which the extreme element with a null measure that initiates the process is not a part of the boundary of Ω. We will once again consider problem $(\mathcal{P}_0)$. We consider, for example, the family $\hat{\Omega}_s =]a/2 - s, a/2 + s[\times\mathcal{O}$ of sub-cylinders built symmetrically around the section $\Gamma_{a/2}$ and bounded by $\Gamma_{s-} = \{a/2 - s\} \times \mathcal{O}$ and $\Gamma_{s+} = \{a/2 + s\} \times \mathcal{O}$. Therefore, we will define the family of problems:

$$(\hat{\mathcal{P}}_{s,h,k}) \begin{cases} -\Delta \hat{u}_{s,h,k} = f \ \text{ in } \hat{\Omega}_s, \\ \hat{u}_{s,h,k}|_\Sigma = 0, \quad -\dfrac{\partial \hat{u}_{s,h,k}}{\partial x}|_{\Gamma_{s-}} = h, \quad \hat{u}_{s,h,k}|_{\Gamma_{s+}} = k, \end{cases}$$

that depends on two boundary data $h \in H^{1/2}_{0,0}(\mathcal{O})'$, $k \in H^{1/2}_{0,0}(\mathcal{O})$. To simplify the notation, we set:

$$\hat{u}_s = \hat{u}_{s,h,k}.$$

Recalling the argument of linearity used previously, we can define the relation between the given data h and k, and the observations $\hat{u}_s$ on Γ_{s-} and $\dfrac{\partial \hat{u}_s}{\partial x}$ on Γ_{s+} by:

$$\hat{u}_s(a/2 - s) = \hat{Q}(s)h + R(s)k + r_1(s),$$

$$\frac{\mathrm{d}\hat{u}_s}{\mathrm{d}x}(a/2 + s) = S(s)h + \hat{P}(s)k + r_2(s),$$

where $\hat{Q} \in \mathcal{L}(H^{1/2}_{0,0}(\mathcal{O})', H^{1/2}_{0,0}(\mathcal{O}))$, $\hat{P} \in \mathcal{L}(H^{1/2}_{0,0}(\mathcal{O}), H^{1/2}_{0,0}(\mathcal{O})')$, $R \in \mathcal{L}(H^{1/2}_{0,0}(\mathcal{O}), H^{1/2}_{0,0}(\mathcal{O}))$, $S \in \mathcal{L}(H^{1/2}_{0,0}(\mathcal{O})', H^{1/2}_{0,0}(\mathcal{O})')$. Following the reasoning used in proposition 1.2, we define the operator:

$$\Pi(s) = \begin{bmatrix} \hat{Q}(s)^\star & 0 \\ R(s)^\star + S(s) & \hat{P}(s) \end{bmatrix}. \qquad [1.51]$$

It is continuous from $H^{1/2}_{0,0}(\mathcal{O})' \times H^{1/2}_{0,0}(\mathcal{O})$ to $H^{1/2}_{0,0}(\mathcal{O}) \times H^{1/2}_{0,0}(\mathcal{O})'$. Furthermore, denoting $\hat{\gamma}_s$ and $\overline{\hat{\gamma}}_s$ to the solutions of $(\hat{\mathcal{P}}_{s,h,k})$ corresponding to

$f = 0$ and to the data h, k and $\overline{h}, \overline{k}$, we have, respectively:

$$\int_{\hat{\Omega}_s} \nabla \hat{\gamma}_s \nabla \overline{\hat{\gamma}}_s \, \mathrm{d}x \, \mathrm{d}z = < h, (\hat{Q}(s)\overline{h} + R(s)\overline{k}) >$$
$$+ < (S(s)h + \hat{P}(s)k), \overline{k} >$$
$$= \Pi(s) \begin{bmatrix} h \\ k \end{bmatrix} \cdot \begin{bmatrix} \overline{h} \\ \overline{k} \end{bmatrix},$$

where $<,>$ represents the duality pairing between $H^{1/2}_{0,0}(\mathcal{O})'$ and $H^{1/2}_{0,0}(\mathcal{O})$. By permuting the data, we deduce that $\Pi(s)$ is self-adjoint, and hence $\hat{P}(s) = \hat{P}(s)^\star$, $\hat{Q}(s) = \hat{Q}(s)^\star$ and $S(s) = -R(s)^\star$. Taking $(h, k) = (\overline{h}, \overline{k})$, we conclude that $\Pi(s)$ is non-negative for $s \in [0, a/2]$. The bilinear form associated with $\Pi(s)$ is coercive on $H^{1/2}_{0,0}(\mathcal{O})' \times H^{1/2}_{0,0}(\mathcal{O})$ for $s \in]0, a/2]$ as a result of the Poincaré inequality and proposition 1.1.

We will now consider the restriction of $\hat{u}_s(h, k)$ to $\hat{\Omega}_x$, for arbitrary h and k, and $x \leq s$. These restrictions satisfy $(\hat{\mathcal{P}}_{x, -\frac{\partial \hat{u}_s(h,k)}{\partial x}|_{\Gamma_{x-}}, \hat{u}_s(h,k)|_{\Gamma_{x+}}})$ on the interval $]a/2 - x, a/2 + x[$, from which:

$$\hat{u}_s(a/2 - x) = -\hat{Q}(x) \frac{\mathrm{d}\hat{u}_s}{\mathrm{d}x}(a/2 - x) + R(x)\hat{u}_s(a/2 + x) + r_1(x), \quad [1.52]$$
$$\frac{\mathrm{d}\hat{u}_s}{\mathrm{d}x}(a/2 + x) = R(x)^\star \frac{\mathrm{d}\hat{u}_s}{\mathrm{d}x}(a/2 - x) + \hat{P}(x)\hat{u}_s(a/2 + x) + r_2(x). \quad [1.53]$$

Formally taking the derivative of [1.52] with respect to x, using the equation satisfied by $\hat{u}_s$,

$$\frac{\partial^2 \hat{u}_s}{\partial x^2} = -\Delta_y \hat{u}_s - f,$$

and substituting $\hat{u}_s(a/2 - x)$ and $\frac{\partial \hat{u}_s}{\partial x}(a/2 + x)$ from [1.52] and [1.53], we obtain:

$$\left(\frac{\mathrm{d}\hat{Q}}{\mathrm{d}x} - \hat{Q}\Delta_y \hat{Q} - RR^\star - I \right) \frac{\partial \hat{u}_s}{\partial x}(a/2 - x)$$
$$- \left(\frac{\mathrm{d}R}{\mathrm{d}x} + R\hat{P} - \hat{Q}\Delta_y R \right) \hat{u}_s(a/2 + x)$$
$$- \frac{\mathrm{d}r_1}{\mathrm{d}x} + \hat{Q}\Delta_y r_1 - Rr_2 + \hat{Q}f(a/2 - x) = 0.$$

We set $x = s$. Using the fact that h and k are arbitrary, we deduce from the previous relation:

$$\frac{\mathrm{d}\hat{Q}}{\mathrm{d}x} - \hat{Q}\Delta_y\hat{Q} - RR^\star = I, \qquad [1.54]$$

$$\frac{\mathrm{d}R}{\mathrm{d}x} + R\hat{P} - \hat{Q}\Delta_y R = 0, \qquad [1.55]$$

on the interval $x \in]0, a/2[$ and for the residual term:

$$\frac{\mathrm{d}r_1}{\mathrm{d}x} - \hat{Q}\Delta_y r_1 + Rr_2 = \hat{Q}f(a/2 - x). \qquad [1.56]$$

Using the same reasoning, from [1.53], we obtain:

$$\frac{\mathrm{d}R^\star}{\mathrm{d}x} + \hat{P}R^\star - R^\star\Delta_y\hat{Q} = 0, \qquad [1.57]$$

$$\frac{\mathrm{d}\hat{P}}{\mathrm{d}x} + \hat{P}^2 + R^\star\Delta_y R + \Delta_y = 0, \qquad [1.58]$$

where [1.57] is equivalent to [1.55]. The residual term r_2 satisfies:

$$\frac{\mathrm{d}r_2}{\mathrm{d}x} + \hat{P}r_2 + R^\star\Delta_y r_1 = -f(a/2 + x) - R^\star f(a/2 - x). \qquad [1.59]$$

Using the arbitrary trajectory argument and the continuity of u in $H^{1/2}_{0,0}(\mathcal{O})$ and of $\frac{\partial u}{\partial x}$ in $H^{1/2}_{0,0}(\mathcal{O})'$ *at* $x = a/2$, from [1.52] and [1.53], we obtain:

$$\hat{Q}(0) = 0, \quad \hat{P}(0) = 0, \quad R(0) = I, \qquad [1.60]$$

and for the residual terms:

$$r_1(0) = 0, \quad r_2(0) = 0. \qquad [1.61]$$

We can rewrite this factorization in vector form by setting:

$$\rho(x) = \begin{pmatrix} r_1(x) \\ r_2(x) \end{pmatrix}, \quad \nu(x) = \begin{pmatrix} u(a/2 - x) \\ u(a/2 + x) \end{pmatrix},$$

$$\phi(x) = \begin{pmatrix} f(a/2 - x) \\ f(a/2 + x) \end{pmatrix}, \qquad [1.62]$$

and consequently:

$$\frac{\mathrm{d}\nu}{\mathrm{d}x}(x) = \begin{pmatrix} -\dfrac{\partial u}{\partial x}(a/2 - x) \\ \dfrac{\partial u}{\partial x}(a/2 + x) \end{pmatrix}. \qquad [1.63]$$

Setting:

$$\Theta(x) = \begin{pmatrix} -\hat{Q}(x) & R(x) \\ R^{\star}(x) & \hat{P}(x) \end{pmatrix}, \quad \Lambda(x) = \begin{pmatrix} -\hat{Q}(x) & 0 \\ R^{\star}(x) & I \end{pmatrix},$$

$$\Phi(x) = \begin{pmatrix} -I & R(x) \\ 0 & \hat{P}(x) \end{pmatrix}, \qquad [1.64]$$

the Riccati equation becomes:

$$\frac{\mathrm{d}\Theta}{\mathrm{d}x} + \Theta \begin{pmatrix} \Delta_y & 0 \\ 0 & I \end{pmatrix} \Theta = \begin{pmatrix} -I & 0 \\ 0 & -\Delta_y \end{pmatrix}, \quad \Theta(0) = \begin{pmatrix} 0 & I \\ I & 0 \end{pmatrix}, \qquad [1.65]$$

to be solved on $x \in]0, a/2[$. Once Θ is known, we can deduce Λ and Φ and the decoupled system is written:

$$\frac{\mathrm{d}\rho}{\mathrm{d}x} + \Theta \begin{pmatrix} \Delta_y & 0 \\ 0 & I \end{pmatrix} \rho = -\Lambda\phi, \quad \rho(0) = 0. \qquad [1.66]$$

Then, we solve for x varying from $a/2$ to 0:

$$\Lambda \frac{\mathrm{d}\nu}{\mathrm{d}x} = \Phi\nu + \rho, \quad \nu(a/2) = \begin{pmatrix} \hat{Q}(a/2)u_0 + R(a/2)u_1 \\ u_1 \end{pmatrix}. \qquad [1.67]$$

REMARK 1.11.– A more natural method, consisting of expressing the Neumann conditions on Γ_{s-} and Γ_{s+} as a function of Dirichlet data on the same sections, leads to a singularity in $x = a/2$ when expressing the initial conditions for the operator satisfying the Riccati equation.

REMARK 1.12.– We can also imagine propagating Γ_{s-} and Γ_{s+} from Γ_0 and Γ_a, respectively, moving them towards $\Gamma_{a/2}$. The advantage of the method given above is that it has no singularities regardless of the boundary conditions on Γ_0 and Γ_a.

2

Justification of the Factorization Computation

In this chapter, we present a justification of the computation which gives the decoupled system [1.22–1.24] that is equivalent to problem $(\mathcal{P}_0)$ presented in section 1.1. Here, we present an approach for the study of the Riccati equation in which we first use a Galerkin method to study a finite dimensional problem and then take the limit as the spatial dimension goes to infinity. This approach was directly inspired by that used by J.-L. Lions in [LIO 68a]. More direct approaches have been proposed by Tartar [TAR 74], using a fixed point method, and by Temam [TEM 71] in the framework of Hilbert–Schmidt operators. Bensoussan, Da Prato, Delfour and Mitter [BEN 07] also considered the fixed point method in their book. As the operator P is unbounded, due to the right-hand side $-\Delta_y$ also being unbounded, all these methods cannot be directly applied to [1.22]. This naturally leads to difficulties in defining P^2. This is why we have chosen to adapt Lions's Galerkin method to [1.19] in Chapter 1. The application of this method to [1.40] in Chapter 1 was presented in [HEN 04]. This Riccati equation was studied in the Hilbert–Schmidt framework in [HEN 08] making use of [TEM 71], that is, without referring to the factorization problem. In [BOU 07], this equation was studied directly using a Yosida regularization.

2.1. Functional framework

We will consider the problem defined by the variational formulation [1.5] from Chapter 1. The spaces X, X_0, Y are defined by definition 1.1 in

Chapter 1. In this chapter, we will use more general spaces. We will begin by recalling the construction of a Hilbert scale related to the Hilbert spaces V and H. Following the approach of [LIO 68b], we assume that H, V are two real separable Hilbert spaces such that $V \subset H$ and V is dense in H with a continuous injection. Then, we can identify H' (topological dual of H) with H and with a dense subspace of V'. We then have

$$V \subset H \subset V',$$

where each space is dense in the next and the injections are continuous.

Suppose that the injection $V \subset H$ is compact and let $\Lambda : V \to V'$ be the canonical isomorphism between V and V', that is to say:

$$< \Lambda u, v >_{V' \times V} = (u, v)_V \quad \forall\, u, v \in V.$$

Using the injection $V \subset H$, we define Λ^{-1} as a compact self-adjoint linear operator $H \to H$. Then (see, for example, theorems VI.8 and VI.11 in [BRE 99]), there exists an orthonormal basis $\{w_i\}_{i=1}^{\infty}$ of H consisting of eigenvectors of Λ^{-1} such that:

$$\Lambda w_i = \lambda_i w_i, \quad \forall\, i \in \mathbb{N},$$

with $\lambda_i > 0$ and:

$$\lim_{i \to \infty} \lambda_i = +\infty.$$

Now:

$$\|w_i\|_V^2 = (w_i, w_i)_V = (\Lambda w_i, w_i)_{V' \times V} = (\lambda_i w_i, w_i)_H = \lambda_i \ \forall\, i = 1, \cdots \infty,$$

and therefore:

$$u = \sum_{i=1}^{\infty} u_i w_i \in V \Leftrightarrow \|u\|_V^2 = \sum_{i=1}^{\infty} \lambda_i |u_i|^2 < \infty.$$

Similarly, if $u = \sum_{i=1}^{\infty} u_i w_i \in V$ and $v = \sum_{i=1}^{\infty} v_i w_i \in V$, then:

$$(u, v)_V = \sum_{i=1}^{\infty} \lambda_i u_i v_i.$$

DEFINITION 2.1.– We define the Hilbert scale H^r, $r \in \mathbb{R}$, as follows:

$$u = \sum_{i=1}^{\infty} u_i w_i \in H^r \Leftrightarrow \|u\|_{H^r}^2 = \sum_{i=1}^{\infty} \lambda_i^r |u_i|^2 < \infty.$$

Furthermore, if $u = \sum_{i=1}^{\infty} u_i w_i \in H^r$ and $v = \sum_{i=1}^{\infty} v_i w_i \in H^r$, then

$$(u, v)_{H^r} = \sum_{i=1}^{\infty} \lambda_i^r u_i v_i.$$

With this definition, we have $H = H^0$, $V = H^1$ and $V' = H^{-1}$. Furthermore, $\{\lambda_i^{-r/2} w_i\}_{i=1}^{\infty}$ is an orthonormal basis H^r, for any $r \in \mathbb{R}$ and Λ is an isomorphism from H^r onto H^{r-2}.

Let us now consider our particular case where $H = L^2(\mathcal{O})$ and $V = H_0^1(\mathcal{O})$. Let Λ be $-\Delta_y$, which is an isomorphism from V to V'. We therefore consider the orthonormal basis of H given by the set $\{w_1, ..., w_n, ..\}$:

$$\begin{cases} -\Delta_y w_n = \lambda_n w_n & \text{in } \mathcal{O} \\ w_n = 0 & \text{on } \partial\mathcal{O}, \end{cases}$$

with $\|w_i\|_{L^2(\mathcal{O})} = 1$, for all $i = 1, 2, \cdots$. It has the following properties:

$$\begin{cases} (a)\ (w_n, w_m)_{L^2(\mathcal{O})} = \delta_{n,m}\ \ \forall\, m, n. \\ (b)\ (w_n, w_m)_{H_0^1(\mathcal{O})} = \int_{\mathcal{O}} \nabla_y w_n(y) \nabla_y w_m(y)\, dy = \lambda_n \delta_{n,m}, \\ (c) \{ \sum_{\text{finite}} \mu_j w_j,\ \mu_j \in \mathbb{R}\} \text{ is a dense subset of } V. \end{cases}$$

Ordering the sequence $\{\lambda_1, ..., \lambda_n, ..\}$ non-decreasingly, we can show as before that $\lambda_i > 0$, for all $i = 1, \cdots, +\infty$ and:

$$\lim_{i \to \infty} \lambda_i = +\infty.$$

The eigenvalues are repeated with their multiplicities and the corresponding eigen-subspaces are of finite dimension. In this case, with

$H = L^2(\mathcal{O})$ and $V = H_0^1(\mathcal{O})$, we have $H^2 = H^2(\mathcal{O}) \cap H_0^1(\mathcal{O})$, $H^4 = \{u \in H^4(\mathcal{O}) : u, \Delta_y u \in H_0^1(\mathcal{O})\}$ and, for $r > 0$:

$$H^r = \{u \in H^r(\mathcal{O}) : \Delta^k u \in H_0^1(\mathcal{O}), \text{ for all } k \in \mathbb{N} \text{ such that } 2k + \frac{1}{2} < r\}.$$

By interpolation and duality, we have $H^{1/2} = H_{0,0}^{1/2}(\mathcal{O})$ and $H^{-1/2} = H_{0,0}^{1/2}(\mathcal{O})'$.

DEFINITION 2.2.– For each $s \in (0, a]$ and $r \in \mathbb{R}$, we define $X_s(r) = L^2(0, s; H^r) \cap H^1(0, s; H^{r-1})$ and $X_{s,0}(r) = \{v \in X_s(r) : v|_{\Gamma_s} = 0\}$.

REMARK 2.1.– From the above definitions, we have in particular: $X = X_a(1)$, $X_0 = X_{a,0}(1)$ and $X_s(r) \subset \mathcal{C}([0, s]; H^{r-\frac{1}{2}})$ with a continuous injection.

To ensure regularity of the boundary value problems defined on the section $\mathcal{O}$, we will assume in this chapter that the boundary $\partial\mathcal{O}$ of $\mathcal{O}$ is of class $\mathcal{C}^\infty$.

2.2. Semi-discretization

In Chapter 1, the operator P and the function w were clearly defined, but the equations that they satisfy were obtained only formally. In order to justify these equations, we will use a Galerkin method.

We define V^m as the space generated by $\{w_1, ..., w_m\}$, $X_s^m = H^1(0, s; V^m)$ and $X_{s,0}^m = \{v \in H^1(0, s; V^m) : v|_{\Gamma_s} = 0\}$. We equip X_s^m and $X_{s,0}^m$ with the following norm:

$$\| v \|_m^2 = \int_0^s \sum_{i=1}^m \left(\lambda_i v_i^2(x) + \left| \frac{\mathrm{d}v_i(x)}{\mathrm{d}x} \right|^2 \right) \mathrm{d}x$$

$$\text{for all } v(x, y) = \sum_{i=1}^m v_i(x) w_i(y) \in X_s^m. \qquad [2.1]$$

We can express any function $h \in H^r$ in the basis $\{w_1, ..., w_n, ..\}$ as $h = \sum_{i=1}^{\infty} h_i w_i$, where $h_i = \frac{(h, w_i)_{H^r}}{\lambda^r}$. We can approach h by $h^m = \sum_{i=1}^{m} h_i w_i \in V^m$.

PROPOSITION 2.1.– The spaces $X_s^m, X_{s,0}^m$ are contained in $\mathcal{C}([0,s]; V^m)$ with continuous injections and $\bigcup_{m\in\mathbb{N}} X_s^m$, $\bigcup_{m\in\mathbb{N}} X_{s,0}^m$ are dense sets of $X_s(r)$, $X_{s,0}(r)$, respectively, for every $r \in \mathbb{R}$.

We define the approximation $u^m \in \{v \in X_a^m : v_{|\Gamma_a} = u_1^m\}$ of the solution u of problem $(\mathcal{P}_0)$ [1.5] as the solution of:

$$(\mathcal{P}_0^m) \qquad \int_\Omega (\nabla u^m \nabla v - fv)\, \mathrm{d}x\mathrm{d}y =< u_0, v|_{\Gamma_0} >_{H_{0,0}^{1/2}(\mathcal{O})' \times H_{0,0}^{1/2}(\mathcal{O})} \quad \forall\, v \in X_{a,0}^m. \qquad [2.2]$$

The coordinates $\{c_i(x)\}_{i=1}^m$ of $u^m(x)$ in the basis $\{w_i\}_{i=1}^m$ satisfy the following system of uncoupled two-point boundary value problems:

$$\begin{cases} -\dfrac{\mathrm{d}^2 c_i}{\mathrm{d}x^2}(x) + \lambda_i c_i(x) = f_i(x) \quad x \in]0, a[, \quad i = 1, .., m \\ -\dfrac{\mathrm{d}c_i}{\mathrm{d}x}(0) =< u_0, w_i >_{H_{0,0}^{1/2}(\mathcal{O})' \times H_{0,0}^{1/2}(\mathcal{O})}, \quad c_i(a) = (u_1, w_i)_{L^2(\mathcal{O})}, \end{cases} \qquad [2.3]$$

where $f_i(x) = \int_\mathcal{O} f(x,y) w_i(y) \mathrm{d}y$. We then have $c_i \in H^2(0,a) \subset \mathcal{C}^1(0,a)$ for $i = 1, \ldots, m$ and:

$$\frac{\partial u^m}{\partial x}(x,y) = \sum_{i=1}^{m} \frac{\mathrm{d}c_i}{\mathrm{d}x}(x) w_i(y) \in X_a^m. \qquad [2.4]$$

As in section 1.2, we embed this problem into a family of problems of the same type, depending on s and h. For all $h \in H^r$ ($r \in \mathbb{R}$ arbitrary), we consider $u_s^m = u_{s,h}^m \in \{v \in X_s^m : v_{|\Gamma_s} = h^m\}$, which is the semi-discrete approximation of $u_{s,h}$ given by the solution of the semi-discrete approximation

$(\mathcal{P}_{s,h})$ defined by:

$$(\mathcal{P}^m_{s,h}) \qquad \int_{\Omega_s} (\nabla u^m_s \nabla v - fv)\,dxdy =< u_0, v|_{\Gamma_0} >_{H^{1/2}_{0,0}{}' \times H^{1/2}_{0,0}} \quad \forall\, v \in X^m_{s,0}. \qquad [2.5]$$

It is clear that $(\mathcal{P}^m_0)$ is exactly $(\mathcal{P}^m_{s,h})$ for $s = a$, $h = u_1$, and then $u^m = u^m_{a,u_1}$. We denote by $\gamma^m_s \in \{v \in X^m_s : v_{|\Gamma_s} = h^m\}$ the part of u^m_s that is linearly dependent on h (to simplify the notation, we do not write the dependency on h), and $\beta^m_s \in X^m_{s,0}$ as the part of u^m_s independent of h and $f^m(x,y) = \sum_{i=1}^m f_i(x)w_i(y)$. They satisfy:

$$\int_0^s\int_{\mathcal{O}} \nabla\gamma^m_s \nabla v\,dxdy = 0 \;\; \forall\, v \in X^m_{s,0}, \qquad [2.6]$$

$$\int_0^s\int_{\mathcal{O}} \nabla\beta^m_s \nabla v\,dxdy = \int_0^s\int_{\mathcal{O}} f^m v\,dxdy + < u_0, v(0) >_{H^{1/2}_{0,0}{}' \times H^{1/2}_{0,0}} \quad \forall\, v \in X^m_{s,0}.$$

Let us then define the linear operator $P^m(s) : H^r \to V^m$ (the same operator for all $r \in \mathbb{R}$) by: $P^m(s)h = P^m(s)h^m = \dfrac{\partial\gamma^m_s}{\partial x}|_{\Gamma_s}$ and let us fix the initial condition $P^m(0) = 0$. We define $w^m(s) = \dfrac{\partial\beta^m_s}{\partial x}|_{\Gamma_s}$ and $w^m(0) = -u_0$. Then:

$P^m(s) \in \mathcal{L}(V^m, V^m)$ and $w^m(s) \in V^m$.

By linearity of $(\mathcal{P}^m_{s,h})$, we have:

$$\frac{du^m_{s,h}}{dx}(s) = P^m(s)u^m_{s,h}(s) + w^m(s).$$

The operator $P^m(s)$ has the same properties as $P(s)$ given in proposition 1.2. We now consider the restriction of u^m_s to Ω_x for $x \le s$. This is the solution of problem $(\mathcal{P}^m_{x,u^m_{s,h}(x)})$. By the same reasoning as above, we have the relation:

$$\frac{du^m_{s,h}}{dx}(x) = P^m(x)u^m_{s,h}(x) + w^m(x). \qquad [2.7]$$

We will assume for now that $P^m(x)$ and $w^m(x)$ are differentiable. Differentiating [2.7] and taking the scalar product with w_j in $L^2(\mathcal{O})$, which we denote by $(.,.)$, we obtain:

$$\left(\frac{\mathrm{d}^2 u_s^m}{\mathrm{d}x^2}, w_j\right) = \left(\frac{\mathrm{d}P^m}{\mathrm{d}x} u_s^m, w_j\right) + \left(P^m \frac{\mathrm{d}u_s^m}{\mathrm{d}x}, w_j\right) + \left(\frac{\mathrm{d}w^m}{\mathrm{d}x}, w_j\right). \quad [2.8]$$

The left-hand side can be evaluated using [2.3]:

$$\left(\frac{\mathrm{d}^2 u_s^m}{\mathrm{d}x^2}, w_j\right) = \left(\sum_{i=1}^{m} ([\lambda_i c_{s,i}(h) - (f, w_i)]w_i), w_j\right)$$
$$=(\Lambda^m u_s^m, w_j) - (f^m, w_j),$$

where Λ^m is the diagonal matrix of dimension $m \times m$ whose diagonal elements are $\{\lambda_i\}_{i=1}^m$. This matrix represents the operator $-\Delta_y$ in the basis $\{w_i\}_{i=1}^m$ of V^m. Furthermore, using [2.7]:

$$\left(P^m \frac{\mathrm{d}u_s^m}{\mathrm{d}x}, w_j\right) = \left((P^m)^2 u_s^m + P^m w^m, w_j\right),$$

which we substitute in [2.8], we obtain:

$$0 = \left(\left[\frac{\mathrm{d}P^m}{\mathrm{d}x}(x) + (P^m)^2(x) + \Delta_y\right] u_s^m(x) + \right.$$
$$\left. + \frac{\mathrm{d}w^m}{\mathrm{d}x}(x) + P^m(x) w^m(x) + f^m(x), \, w_j\right), \quad [2.9]$$

for $j \in \{1, \cdots, m\}$ and $0 \leq x \leq s$. Setting $x = s$ and using the fact that h and s are arbitrary, we can deduce from the previous identity that:

$$\begin{cases} \frac{\mathrm{d}P^m}{\mathrm{d}x}(x) + (P^m)^2(x) + \Delta_y = 0, \; 0 < x \leq a, \\ P^m(0) = 0, \end{cases} \quad [2.10]$$

$$\begin{cases} \frac{\mathrm{d}w^m}{\mathrm{d}x}(x) + P^m(x) w^m(x) = -f^m(x), \; 0 < x \leq a, \\ w^m(0) = -u_0^m. \end{cases} \quad [2.11]$$

We still have to justify the differentiability of P^m and w^m. From the theory of ordinary differential equations, we know that there exists a local solution

P^m of [2.10] in $[0, \delta]$, for sufficiently small δ. Furthermore, P^m is $\mathcal{C}^1$ from $[0, \delta]$ to $\mathcal{L}(V^m, V^m)$. From this, we deduce (see, for example, theorem 2 in [KAT 75]), for the case of general functions f, that w^m is a solution of [2.11] in $[0, \delta]$ and that $w^m \in H^1(0, \delta; V^m)$. Furthermore, for these solutions and $s \in [0, \delta]$, [2.7] is satisfied. Before, we can take this any further, we need estimates of $P^m(s)$ that are independent of s.

PROPOSITION 2.2.– The operator P^m is uniformly bounded in m in $L^\infty\left(0, a;\ \mathcal{L}\left(H^r, H^{r-1}\right)\right)$, for all $r \in \mathbb{R}$. Furthermore w^m is uniformly bounded in $L^\infty(0, a; H^{1/2}_{0,0}(\mathcal{O})')$.

PROOF.– Let $v(y) = \sum_{i=1}^m v_i w_i(y) \in V^m$. For each $r \in \mathbb{R}$, we define the following norm on V^m:

$$\| v \|_{m,r}^2 = \sum_{i=1}^{m} (\lambda_i)^r v_i^2,$$

which is uniformly equivalent in m to the norm of $H^r(\mathcal{O})$ (see definition 2.1).

Let $s \in (0, a]$ and $h = \sum_{i=1}^{\infty} h_i w_i \in H^r$. Then, $P^m(s)h = \frac{\partial \gamma_s^m}{\partial x}|_{\Gamma_s}$. We consider $\gamma_s^m(x) = \sum_{i=1}^m c_i(x) w_i$, where $\{c_i\}_{i=1}^m$ is the solution of [2.3] for $x \in [0, s]$ with $f_i = 0$, $u_0 = 0$ and $u_1 = h$. We have:

$$\begin{aligned} \left|\frac{\mathrm{d}c_i}{\mathrm{d}x}(s)\right|^2 &= \int_0^s \frac{\mathrm{d}}{\mathrm{d}x}\left(\frac{\mathrm{d}c_i}{\mathrm{d}x}(x)\right)^2 \mathrm{d}x \\ &\leq 2\left(\int_0^s \left|\frac{\mathrm{d}c_i}{\mathrm{d}x}(x)\right|^2 \mathrm{d}x\right)^{1/2} \left(\int_0^s \lambda_i^2 \left|c_i(x)\right|^2 \mathrm{d}x\right)^{1/2}. \end{aligned} \quad [2.12]$$

Multiplying [2.12] by $\lambda_i^{r-1} = \lambda_i^{r/2-1/4} \lambda_i^{r/2-3/4}$, summing over i and using the Cauchy–Schwartz inequality, we obtain:

$$\begin{aligned} \| \frac{\partial \gamma_s^m}{\partial x}|_{\Gamma_s} \|_{m,r-1}^2 \leq 2 &\left(\int_0^s \left\|\frac{\partial \gamma_s^m}{\partial x}\right\|_{m,r-1/2}^2 \mathrm{d}x\right)^{1/2} \\ &\left(\int_0^s \| \gamma_s^m \|_{m,r+1/2}^2 \mathrm{d}x\right)^{1/2}. \end{aligned} \quad [2.13]$$

Now, considering [2.3] with $a = s$, $f_i = 0$, $u_0 = 0$ and $u_1 = h$, multiplying each equation by $(\lambda_i)^{r-1/2} c_i$, summing over i, integrating w.r.t. x and then integrating by parts, we obtain:

$$\int_0^s \left\| \frac{\partial \gamma_s^m}{\partial x} \right\|_{m,r-1/2}^2 \mathrm{d}x + \int_0^s \| \gamma_s^m \|_{m,r+1/2}^2 \,\mathrm{d}x = \sum_{i=1}^{m} (\lambda_i)^{r-1/2} h_i \frac{\mathrm{d}c_i}{\mathrm{d}x}(s)$$

$$\leq \| h \|_{m,r} \| \frac{\partial \gamma_s^m}{\partial x} |_{\Gamma_s} \|_{m,r-1} . \quad [2.14]$$

Using Young's inequality and estimates [2.13] and [2.14], we obtain:

$$\| P^m(s) h \|_{m,r-1} = \| \frac{\partial \gamma_s^m}{\partial x} |_{\Gamma_s} \|_{m,r-1} \leq \| h \|_{m,r}, \quad [2.15]$$

which proves that P^m is uniformly bounded in m, in $L^\infty\left(0, a; \mathcal{L}\left(H^r, H^{r-1}\right)\right)$.

The proof for r^m using estimates of β_s^m is similar. □

COROLLARY 2.1.– The operator P^m is uniformly bounded in m, in the following spaces:
$L^\infty\left(0, a; \mathcal{L}\left(H_{0,0}^{1/2}(\mathcal{O}), H_{0,0}^{1/2}(\mathcal{O})'\right)\right)$,
$L^\infty\left(0, a; \mathcal{L}\left(H_0^1(\mathcal{O}) \cap H^{3/2}(\mathcal{O}), H_{0,0}^{1/2}(\mathcal{O})\right)\right)$,
$L^\infty\left(0, a; \mathcal{L}\left(H_0^1(\mathcal{O}), L^2(\mathcal{O})\right)\right)$
and $L^\infty\left(0, a; \mathcal{L}\left(L^2(\mathcal{O}), H^{-1}(\mathcal{O})\right)\right)$.

PROOF.– The proof can be obtained by using proposition 2.2 and the fact that $H^0 = L^2(\mathcal{O})$, $H^1 = H_0^1(\mathcal{O})$, $H^{1/2} = H_{0,0}^{1/2}(\mathcal{O})$, $H^{-1/2} = H_{0,0}^{1/2}(\mathcal{O})'$, $H^{3/2} = H_0^1(\mathcal{O}) \cap H^{3/2}(\mathcal{O})$ and $H^{-1} = H^{-1}(\mathcal{O})$ (see [LIO 68b]). □

REMARK 2.2.– By proposition 2.2, we deduce that P^m is a global solution of [2.10] and $\mathcal{C}^1$ from $[0, a] \to \mathcal{L}(V^m, V^m)$. Hence, applying once again theorem 2 from [KAT 75], we find that w^m is a global solution of [2.11] and that $w^m \in H^1(0, a; V^m)$.

2.3. Passing to the limit

We will now study the convergence of u^m as m goes to infinity.

THEOREM 2.1.– Let $s \in]0,a]$ and $h = \sum_{i=1}^{\infty} h_i w_i \in H^r$ be arbitrary, $h^m = \sum_{i=1}^{m} h_i w_i \in V^m$, and let γ_s, γ_s^m be the solutions of [1.7] from Chapter 1 and [2.6], respectively. As $m \to \infty$, we have $\gamma_s^m \to \gamma_s$ in $X_s(r+1/2)$, $\gamma_s^m(x) \to \gamma_s(x)$ in H^r and $\frac{\mathrm{d}\gamma_s^m}{\mathrm{d}x}(x) \to \frac{\mathrm{d}\gamma_s}{\mathrm{d}x}(x)$ in H^{r-1}, for every $x \in [0,s]$.

PROOF.– From [2.3], with $f = 0$, $a = s$, $u_0 = 0$ and $u_1 = h$, we have:

$$\int_0^s \left(\left| \frac{\mathrm{d}c_i}{\mathrm{d}x}(x) \right|^2 + \lambda_i |c_i(x)|^2 \right) \mathrm{d}x = h_i \frac{\mathrm{d}c_i}{\mathrm{d}x}(s).$$

Multiplying by $(\lambda_i)^{r-1/2}$, we have:

$$\int_0^s \left((\lambda_i)^{r-1/2} \left| \frac{\mathrm{d}c_i}{\mathrm{d}x}(x) \right|^2 + (\lambda_i)^{r+1/2} |c_i(x)|^2 \right) \mathrm{d}x = \\ = (\lambda_i)^{r-1/2} h_i \frac{\mathrm{d}c_i}{\mathrm{d}x}(s). \quad [2.16]$$

Summing over i and applying the Cauchy–Schwarz inequality, we obtain:

$$\begin{aligned} \|\gamma_s^m\|^2_{X_s(r+1/2)} &= \sum_{i=1}^{m} \int_0^s \left((\lambda_i)^{r-1/2} \left| \frac{\mathrm{d}c_i}{\mathrm{d}x}(x) \right|^2 + (\lambda_i)^{r+1/2} |c_i(x)|^2 \right) \mathrm{d}x \\ &\leq \|h\|_{H^r} \, \|\frac{\mathrm{d}\gamma_s^m}{\mathrm{d}x}(s)\|_{H^{r-1}} \\ &\leq C\|h\|_{H^r} \, \|\gamma_s^m\|_{X_s(r+1/2)}. \end{aligned}$$

Using Young's inequality, we show that $\|\gamma_s^m\|_{X_s(r+1/2)}$ is uniformly bounded in m. Then, if $z = \sum_{i=1}^{\infty} c_i w_i$, we have:

$$\gamma_s^m = \sum_{i=1}^{m} c_i w_i \to z \text{ strongly in } X_s(r+1/2).$$

Furthermore, by construction:

$$-\Delta z = -\frac{\partial^2 z}{\partial x^2} - \Delta_y z = \sum_{i=1}^{\infty} \left(-\frac{\mathrm{d}^2 c_i}{\mathrm{d}x^2} w_i + \lambda_i c_i w_i \right) = 0 \text{ in } X_s(r - 3/2),$$

$$-\frac{\partial z}{\partial x}|_{\Gamma_0} = -\sum_{i=1}^{\infty} \frac{\mathrm{d}c_i}{\mathrm{d}x}(0) w_i = 0$$

and:

$$z|_{\Gamma_s} = \sum_{i=1}^{\infty} c_i(s) w_i = h.$$

Hence, $z = \gamma_s$. Finally, the pointwise convergences follow from remark 2.1. □

COROLLARY 2.2.– If $h \in H^r$, with arbitrary $r \in \mathbb{R}$, then $P^m(s)h \to P(s)h$ strongly in H^{r-1} and for all $s \in [0, a]$.

We have thus uniquely defined $P(s) \in \mathcal{L}(H^r, H^{r-1})$.

THEOREM 2.2.– Let $s \in]0, a]$. As $m \to \infty$, we have $\beta_s^m \to \beta_s$ in $X_s(1)$, $\beta_s^m(x) \to \beta_s(x)$ in $H^{1/2} = H_{0,0}^{1/2}(\mathcal{O})$ and $\frac{\mathrm{d}\beta_s^m}{\mathrm{d}x}(x) \to \frac{\mathrm{d}\beta_s}{\mathrm{d}x}(x)$ in $H^{-1/2} = H_{0,0}^{1/2}(\mathcal{O})'$, for every $x \in [0, a]$.

PROOF.– From [2.3], with $a = s$ and $u_1 = 0$, we have:

$$\int_0^s \left(\left| \frac{\mathrm{d}c_i}{\mathrm{d}x}(x) \right|^2 + \lambda_i |c_i(x)|^2 \right) \mathrm{d}x = \int_0^s f_i(x) c_i(x) \, \mathrm{d}x$$
$$+ < u_0, w_i >_{H_{0,0}^{1/2}(\mathcal{O})' \times H_{0,0}^{1/2}(\mathcal{O})} c_i(0).$$

Summing over i, applying the Cauchy–Schwarz inequality and using continuity of the injection $X_s(1) \subset C([0, s]; H^{1/2})$ (see remark 2.1), we obtain:

$$\int_{\Omega_s} |\nabla \beta_s^m|^2 \mathrm{d}x\mathrm{d}y \leq \| f \|_{L^2(0,s;H^{-1}(\mathcal{O}))} \| \beta_s^m \|_{L^2(0,s;H_0^1(\mathcal{O}))}$$
$$+ \| u_0 \|_{H_{0,0}^{1/2}(\mathcal{O})'} \| \beta_s(0) \|_{H_{0,0}^{1/2}(\mathcal{O})}$$
$$\leq \left(\| f \|_{L^2(0,s;H^{-1}(\mathcal{O}))} + C \| u_0 \|_{H_{0,0}^{1/2}(\mathcal{O})'} \right) \| \beta_s^m \|_{H^1(\Omega_s)} .$$

Then, using Young's inequality, we show that $\|\beta_s^m\|_{X_s(1)} = \|\beta_s^m\|_{H^1(\Omega_s)}$ is uniformly bounded in m. Consequently:

$$\beta_s^m = \sum_{i=1}^{m} c_i w_i \to z = \sum_{i=1}^{\infty} c_i w_i \text{ strongly in } X_s(1).$$

Furthermore, by construction:

$$-\Delta z = -\frac{\partial^2 z}{\partial x^2} - \Delta_y z = \sum_{i=1}^{\infty} \left(-\frac{\mathrm{d}^2 c_i}{\mathrm{d}x^2} w_i + \lambda_i c_i w_i \right) = \sum_{i=1}^{\infty} f_i w_i = f$$
$$\text{in } L^2(0,s;H^{-1}(\mathcal{O})),$$

$$-\frac{\partial z}{\partial x}|_{\Gamma_0} = -\sum_{i=1}^{\infty} \frac{\mathrm{d}c_i}{\mathrm{d}x}(0) w_i = \sum_{i=1}^{\infty} < u_0, w_i >_{H_{0,0}^{1/2}(\mathcal{O})' \times H_{0,0}^{1/2}(\mathcal{O})} w_i = u_0$$

and

$$z|_{\Gamma_s} = \sum_{i=1}^{\infty} c_i(s) w_i = 0.$$

Then, $z = \beta_s$. Finally, pointwise convergence follows from proposition 1.1 in Chapter 1. □

REMARK 2.3.– If f and u_0 are more regular, then β_s will be more regular and so will the convergence of β_s^m.

COROLLARY 2.3.– As $m \to \infty$, we have $w^m(s) \to w(s)$ in $H_{0,0}^{1/2}(\mathcal{O})'$ $\forall s \in]0,a]$.

COROLLARY 2.4.– As $m \to \infty$, we have $u^m \to u$ in $X_a(1)$, $u^m(x) \to u(x)$ in $H^{1/2} = H^{1/2}_{0,0}(\mathcal{O})$ and $\dfrac{du^m}{dx}(x) \to \dfrac{du}{dx}(x)$ in $H^{-1/2} = H^{1/2}_{0,0}(\mathcal{O})'$, for all $x \in [0, a]$.

COROLLARY 2.5.– The function w is the solution of the following system in $X_a(0) = L^2(0, a; L^2(\mathcal{O})) \cap H^1(0, a; H^{-1}(\mathcal{O})) \subset \mathcal{C}([0, a]; H^{1/2}_{0,0}(\mathcal{O})')$:

$$\begin{cases} \dfrac{dw}{dx}(x) + P(x)w(x) = -f(x), & \text{in } L^2(0, a; H^{-1}(\mathcal{O})), \\ w(0) = -u_0, \end{cases} \qquad [2.17]$$

and we have the convergence $w^m \to w$ in $X_a(0)$.

PROOF.– Applying corollaries 2.2, 2.3 and 2.4, for every $s \in]0, a]$, we obtain:

$$\frac{du^m}{dx}(s) = P^m(s)u^m(s) + w^m(s) \to \frac{du}{dx}(s) = P(s)u(s) + w(s)$$
$$\text{strongly in } H^{1/2}_{0,0}(\mathcal{O})'. \qquad [2.18]$$

Given that:

$$u \in L^2(0, a; H^1_0(\mathcal{O})) \cap H^1(0, a; L^2(\mathcal{O})) \subset \mathcal{C}([0, a]; L^2(\mathcal{O}))$$

and $P \in L^\infty\left(0, a; \mathcal{L}\left(H^1_0(\mathcal{O}), L^2(\mathcal{O})\right)\right)$, we have that $P(x)u(x) \in L^2(0, a; L^2(\mathcal{O}))$ and:

$$w(x) = \frac{du}{dx}(x) - P(x)u(x) \in L^2(0, a; L^2(\mathcal{O})).$$

Now, as:

$$\frac{dw^m}{dx}(x) + P^m(x)w^m(x) = -f^m(x),\ 0 < x \leq a$$

and using corollary 2.1 $P \in L^\infty\left(0, a; \mathcal{L}\left(L^2(\mathcal{O}), H^{-1}(\mathcal{O})\right)\right)$, we have that:

$$\frac{dw^m}{dx}(x) = -P^m(x)w^m(x) - f^m(x)$$

is uniformly bounded in $L^2(0,a;H^{-1}(\mathcal{O}))$. Furthermore, applying corollary 2.2, we can take the limit:

$$\frac{dw}{dx}(x) = -P(x)w(x) - f(x) \text{ in } L^2(0,a : H^{-1}(\mathcal{O})).$$

This shows that:

$$w \in L^2(0,a;L^2(\mathcal{O})) \cap H^1(0,a;H^{-1}(\mathcal{O})) \subset \mathcal{C}([0,a];H^{1/2}_{0,0}(\mathcal{O})').$$

The initial condition is also satisfied because:

$$w^m(0) = \frac{du^m}{dx}(0) - P^m(0)u(0) = \sum_{i=1}^{m} - < u_0, w_i >_{H^{1/2}_{0,0}(\mathcal{O})' \times H^{1/2}_{0,0}(\mathcal{O})} w_i$$

$$\to u_0 \text{ in } H^{1/2}_{0,0}(\mathcal{O})'. \qquad [2.19]$$

□

PROPOSITION 2.3.– Let $r \in \mathbb{R}$ and $s \in [0,a]$. Then, $P(s) : H^r \mapsto H^{r-1}$ is a self-adjoint operator. The associated bilinear form:

$$< P(s)h, h >_{H^{r-1} \times H^r},$$

where $< \cdot, \cdot >_{H^{r-1} \times H^r}$ is the duality pairing with respect to the pivot space $H^{r-1/2}$, is coercive in H^r for $s \in]0,a]$. There exists a constant $C > 0$, independent of a, such that:

$$\frac{s}{C+s}$$

is a constant of coercivity for this bilinear form. Furthermore, $\|P(s)\|_{\mathcal{L}(H^r,H^{r-1})}$ is bounded independently of r. Finally, $P(s)$ and $P(x)$ commute in the following sense:

$$(P(x)h, P(s)\bar{h})_{H^{r-1}} = (P(s)h, P(x)\bar{h})_{H^{r-1}}, \quad \forall h, \bar{h} \in H^r.$$

PROOF.– From [2.3], with $f = 0$, $a = s$, $u_0 = 0$ and $u_1 = h$, we have:

$$P(s)h = \sum_{i=1}^{\infty} \frac{dc_i}{dx}(s) w_i$$

and [2.16]. Then, summing over i:

$$\|\gamma_s\|^2_{X_s(r+1/2)} = \sum_{i=1}^{\infty} (\lambda_i)^{r-1/2} h_i \frac{\mathrm{d}c_i}{\mathrm{d}x}(s) =< P(x)h, h >_{H^{r-1}\times H^r},$$

showing the positivity of P. Repeating the steps of the proof of proposition 1.2, we have:

$$\|\gamma_s\|^2_{X_s(r+1/2)} \geq \frac{s}{C+s} \|h\|^2_{H^r},$$

which proves the result. The independence w.r.t. r of the upper bound for the norm of P follows from estimate [2.15]. The commutation property is shown by:

$$(P(x)h, P(s)\bar{h})_{H^{r-1}} = \sum_{i=1}^{\infty} (\lambda_i)^{r-1} \frac{\mathrm{d}c_i}{\mathrm{d}x}(x) \frac{\mathrm{d}\bar{c}_i}{\mathrm{d}x}(s),$$

but, because of the homogeneity conditions ($f = 0$, $u_0 = 0$), c_i depends linearly on h_i:

$$c_i(x) = c_i^1(x) h_i.$$

Then:

$$(P(x)h, P(s)\bar{h})_{H^{r-1}} = \sum_{i=1}^{\infty} (\lambda_i)^{r-1} \frac{\mathrm{d}c_i^1}{\mathrm{d}x}(x) \frac{\mathrm{d}c_i^1}{\mathrm{d}x}(s) h_i \bar{h}_i = (P(s)h, P(x)\bar{h})_{H^{r-1}}.$$

□

We will now study the solution of problem [2.17] and show that it is unique.

THEOREM 2.3.– Problem [2.17] has a unique solution in $X_a(0)$.

PROOF.– In corollary 2.5, we had already established the existence of a solution. For the uniqueness, let $w_1, w_2 \in X_a(0)$ be two solutions of [2.17]. Then, for each $s \in [0, a]$, we have:

$$\frac{1}{2}\|w_1(s) - w_2(s)\|^2_{H^{-1/2}} + \int_0^s < P(x)(w_1(x) - w_2(x)), \\ w_1(x) - w_2(x) >_{H^{-1}\times H^0} \mathrm{d}x = 0, \qquad [2.20]$$

where $\|w_1(s) - w_2(s)\|^2_{H^{-1/2}} = 0$ for all $s \in [0,a]$ by the positivity of P shown in proposition 2.3, which proves the uniqueness of the solution. □

REMARK 2.4.– In light of proposition 2.3, we could have shown the existence and uniqueness of the solution of problem [2.17] using general theorems on abstract parabolic problems if P does not vanish at 0. However, we must bear in mind that P is not a differential operator.

Similarly, using corollary 2.5, we can demonstrate the following result for the function u.

THEOREM 2.4.– The solution u of problem $(\mathcal{P}_0)$ from chapter 1 is also a solution of:

$$\frac{du}{dx}(x) = P(x)u(x) + w(x), \qquad u(a) = u_1. \qquad [2.21]$$

Problem [2.21] has a unique solution in $X_a(1)$ for an operator P satisfying the following properties: $P \in L^\infty\left(0,a;\mathcal{L}\left(H^1,H^0\right)\right)$ and $< P(s)u, v >_{H^0 \times H^1} \geq 0$ for every $u, v \in H^1$ and for almost all $s \in [0,a]$.

We can now take the limit in [2.10].

THEOREM 2.5.– The operator P satisfies a Riccati equation in the following sense: for all $h, \overline{h} \in H^r$, for $r \in \mathbb{R}$:

$$\frac{\mathrm{d}}{\mathrm{d}x}(P(x)h,\overline{h})_{H^{r-1}} + (P(x)h, P(x)\overline{h})_{H^{r-1}} =$$
$$= (\nabla_y h, \nabla_y \overline{h})_{H^{r-1}} \in \mathcal{D}'(]0,a[) \quad [2.22]$$

with the initial condition $P(0) = 0$. Furthermore, $P \in L^\infty\left(0,a;\mathcal{L}\left(H^r,H^{r-1}\right)\right)$ is weakly continuous in x, meaning that $(P(x)h,\overline{h})_{H^{r-1}}$ is continuous on $[0,a]$ $\forall\, h, \overline{h} \in H^r$.

PROOF.– Let $h, \overline{h} \in H^r$. By [2.10], we have, by taking the scalar products in H^{r-1}:

$$\left(\frac{\mathrm{d}P^m}{\mathrm{d}x}(x)h,\overline{h}\right)_{H^{r-1}} + (P^m(x)h, P^m(x)\overline{h})_{H^{r-1}} = (\nabla_y h, \nabla_y \overline{h})_{H^{r-1}},$$

and $P^m(0) = 0$. Let $\varphi \in \mathcal{C}^1([0,a])$ such that $\varphi(a) = 0$. We obtain:

$$-\int_0^a (P^m(x)h, \overline{h})_{H^{r-1}}\varphi(x)'\mathrm{d}x$$
$$+\int_0^a (P^m(x)h, P^m(x)\overline{h})_{H^{r-1}}\varphi(x)\mathrm{d}x = \int_0^a (\nabla_y h, \nabla_y \overline{h})_{H^{r-1}}\varphi(x)\mathrm{d}x. \quad [2.23]$$

The integrand of the first term converges towards $(P(x)h, \overline{h})_{H^{r-1}}\varphi(x)'$ by corollary 2.2 and is bounded for $x \in [0,a]$ by applying corollary 2.1. The convergence of the integrand in the second term can be shown similarly. Consequently, by Lebesgue's theorem, we can take the limit in [2.23]. For $\varphi \in \mathcal{D}(]0,a[)$, we obtain [2.22]. Then, using [2.22], $\frac{\mathrm{d}}{\mathrm{d}x}(P(x)h, \overline{h})_{H^{r-1}}$ belongs to $L^\infty(0,a)$ and therefore $(P(x)h, \overline{h})_{H^{r-1}}$ is continuous in x, and for $\varphi \in \mathcal{C}^1([0,a])$ we can integrate [2.23] by parts, which gives us our initial condition $P(0) = 0$. □

It is natural to focus on the solutions of the Riccati equation [2.22] with the properties listed in proposition 2.3. We will first show a result that follows from these properties independently of the Riccati equation.

PROPOSITION 2.4.– Let Q be an operator of $\mathcal{L}(H^r, H^{r-1})$ for all $r \geq r_0$. Let $\{w_i\}_{i=1}^\infty$ be the Hilbert basis of H^0 defined in section 2.1. Let q_{ij} be the components of Qw_i with respect to the basis $\{w_j\}$: $Qw_i = \sum_{j=1}^\infty q_{ij}w_j$. We assume that Q is symmetrical in the sense that $q_{ij} = q_{ji}$ for every $i, j \in \mathbb{N}$ and such that $\|Q\|_{\mathcal{L}(H^r,H^{r-1})} \leq C$ independently of r. Then, Q is block diagonal, that is, $q_{ij} = 0$ if $\lambda_i \neq \lambda_j$.

PROOF.– Indeed, we have:

$$\|Q\|^2_{\mathcal{L}(H^r,H^{r-1})} \geq \sup_i \frac{\|Qw_i\|^2_{H^{r-1}}}{\|w_i\|^2_{H^r}} = \sup_i \lambda_i^{-r} \sum_{j=1}^\infty \lambda_j^{r-1} q_{ij}^2.$$

Then, we have $q_{ij} = 0$ if $\lambda_i < \lambda_j$ because otherwise, since $\lambda_i^{-r}\lambda_j^r \to \infty$ as $r \to \infty$, we would have

$$\|Q\|_{\mathcal{L}(H^r,H^{r-1})} \to \infty, \text{for } r \to \infty,$$

which contradicts our hypothesis. For $\lambda_i > \lambda_j$, the same reasoning and the symmetry $q_{ij} = q_{ji}$ are used. □

Finally, we find the following result for uniqueness.

THEOREM 2.6.– The solution of the Riccati equation [2.22], as defined in theorem 2.5, is unique in the class of positive operators, in the sense that:

$$\langle P(x)h, h\rangle_{H^{r-1}\times H^r} \geq 0, \qquad \forall h \in H^r,$$

that satisfy the hypotheses of proposition 2.4.

PROOF.– Let $\lambda_{[k]}$ be the distinct eigenvalues of Λ: $\{\lambda_{[k]}\}_{k=1}^{\infty} = \{\lambda_i\}_{i=1}^{\infty}$, such that:

$$\lambda_1 = \lambda_{[1]} < \lambda_{[2]} < \cdots$$

We consider the eigen-subspaces $V_{[k]}$ generated by $w_i, i = 1, \cdots, \infty$, for which $\lambda_i = \lambda_{[k]}$. We can then uniquely write $h \in H^r$ as:

$$h = \sum_{k=1}^{\infty} h_{[k]},$$

with $h_{[k]} \in V_{[k]}$. Let P satisfy the hypotheses of proposition 2.4 and $P_{[k]}$ its restriction on $V_{[k]}$. We have:

$$Ph = \sum_{k=1}^{\infty} P_{[k]} h_{[k]}.$$

Again writing $P_{[k]}$ for its matrix with respect to the basis $\{w_i\}_{i=1}^{\infty}$, this matrix satisfies the matrix Riccati equation:

$$\frac{\mathrm{d}P_{[k]}}{\mathrm{d}x}(x) + (P_{[k]})^2(x) - \lambda_{[k]} I = 0,\ 0 < x \leq a, P_{[k]}(0) = 0.$$

This equation has a unique solution by theorem 4.1.6 in [ABO 03] and therefore so does the solution of [2.22]. □

3

Complements to the Model Problem

In this chapter, we will once again consider the model case presented in Chapter 1 and the method of invariant embedding from section 1.2. We will show that other boundary conditions are possible. Other forms of factorization are also possible to emphasize different aspects of the dependency on the data. Finally, we will discuss the links between this approach by factorization and other classical results from the theory of boundary value problems, such as the representation of solutions by Green's functions or Steklov–Poincaré operators, which are usually used in domain decomposition methods.

3.1. An alternative method for obtaining the factorization

Here, and later, we will require a Green's formula that specifically considers the cylindrical geometry of the domain Ω. In the x-direction, this Green's formula takes the form of integration by parts. It is given by:

$$\int_\Omega -\Delta u\, v \,\mathrm{d}x\,\mathrm{d}y =$$

$$= \int_\Omega \nabla u \nabla v \,\mathrm{d}x\,\mathrm{d}y - \int_\Sigma \frac{\partial u}{\partial \nu} v \,\mathrm{d}\Sigma - \int_{\Gamma_a} \frac{\partial u}{\partial x} v \,\mathrm{d}y + \int_{\Gamma_0} \frac{\partial u}{\partial x} v \,\mathrm{d}y, \quad [3.1]$$

for $u, v \in H^1(\Omega)$ and $\Delta u \in L^2(\Omega)$. This formula may be extended to $\Delta u \in \hat{X}'$ in the notation of definition 1.2. This provides a definition for $\frac{\partial u}{\partial \nu}|_\Sigma$ in $H^{-1/2}(\Sigma)$, and for $\frac{\partial u}{\partial x}|_{\Gamma_a}$ in $H^{-1/2}(\Gamma_a)$ and similarly on Γ_0. We will also use another Green's formula in which the operator is the Laplacian restricted to the sections Δ_y:

$$< u, v >_{\hat{X}' \times \hat{X}} = \int_{\Omega} \nabla_y u \nabla_y v \, \mathrm{d}x \, \mathrm{d}y - \int_{\Sigma} \frac{\partial u}{\partial \nu} v \, \mathrm{d}\Sigma, \qquad [3.2]$$

for $u, v \in L^2(0, a; H^1(\Omega))$. This allows $\frac{\partial u}{\partial \nu}$ to be defined in $L^2(0, a; H^{-1/2}(\partial\Omega))$.

We now return to the question of the factorization of problem $(\mathcal{P}_0)$, presenting a method that is equivalent to the method in section 1.2, but that is based on an application of bilinear forms of the type commonly used in variational formulations. This method is easier to extend to different kinds of boundary conditions.

Consider the family of problems $(\mathcal{P}_{s,h})$ defined in section 1.2, with solutions given by $u_{s,h}$. Let u be the restriction of $u_{s,h}$ to Ω_x for $0 < x \leq s$. Multiplying this equation by a function $\varphi \in H_0^1(\mathcal{O})$ constant in x, then integrating over Ω_x for $x \leq s$ and applying Green's formula [3.1], we obtain:

$$\int_0^x (\nabla_y u, \nabla_y \varphi) \, \mathrm{d}x' = \int_0^x < f, \varphi >_{H^{-1}(\mathcal{O}) \times H_0^1(\mathcal{O})} \, \mathrm{d}x' + \int_{\Gamma_x} \frac{\partial u}{\partial x} \varphi \, \mathrm{d}y$$
$$+ \int_{\Gamma_0} u_0 \varphi \, \mathrm{d}y, \qquad [3.3]$$

for all φ in $H_0^1(\mathcal{O})$, where $(,)$ is the scalar product in $L^2(\mathcal{O})$. By the argument in section 1.2, there is an affine relationship between the Dirichlet and Neumann boundary conditions, which may be expressed in variational form as follows:

$$\left(\frac{\partial u}{\partial x}(x), \varphi \right) = (P(x) u(x), \varphi) + (w(x), \varphi)$$
$$\forall \varphi \in H_0^1(\mathcal{O}), \text{ for a.e. } x \in]0, s[. \qquad [3.4]$$

Thus, substituting [3.4] into [3.3]:

$$\int_0^x (\nabla_y u, \nabla_y \varphi) \, \mathrm{d}x' = \int_0^x < f, \varphi >_{H^{-1}(\mathcal{O}) \times H_0^1(\mathcal{O})} \, \mathrm{d}x'$$
$$+ \int_{\Gamma_x} (Pu + w) \varphi \, \mathrm{d}y + \int_{\Gamma_0} u_0 \varphi \, \mathrm{d}y. \qquad [3.5]$$

By differentiating [3.5] w.r.t. x using the differentiation formulas for domains given in [SOK 92], which have a very simple expression in this geometry as the section Γ_s moves without deformation, we obtain after setting $x = s$:

$$\int_{\Gamma_s} \nabla_y u \nabla_y \varphi \, \mathrm{d}y =< f, \varphi >_{H^{-1}(\mathcal{O}) \times H_0^1(\mathcal{O})} + \int_{\Gamma_s} \frac{\mathrm{d}}{\mathrm{d}x}(Pu + w)\varphi \, \mathrm{d}y. \quad [3.6]$$

By self-adjointness of P (proposition 1.2), the last term may be rewritten as:

$$\begin{aligned}\int_{\Gamma_s} \frac{\mathrm{d}}{\mathrm{d}x}(Pu + w)\varphi \, \mathrm{d}x &= \int_{\Gamma_s} \frac{\mathrm{d}P}{\mathrm{d}x} u\varphi \, \mathrm{d}y + \int_{\Gamma_s} \frac{\mathrm{d}u}{\mathrm{d}x} P\varphi \, \mathrm{d}y + \int_{\Gamma_s} \frac{\mathrm{d}w}{\mathrm{d}x} \varphi \, \mathrm{d}y, \\ &= \int_{\Gamma_s} \frac{\mathrm{d}P}{\mathrm{d}x} u\varphi \, \mathrm{d}y + \int_{\Gamma_s} PuP\varphi \, \mathrm{d}y \qquad [3.7] \\ &+ \int_{\Gamma_s} (\frac{\mathrm{d}w}{\mathrm{d}x} + Pw)\varphi \, \mathrm{d}y.\end{aligned}$$

By substituting this into [3.6], the bilinear terms in u and φ give the Riccati equation [1.26] and the linear term in φ gives the equation of the residual [1.27].

3.2. Other boundary conditions

In the presentation of section 1.2, we chose boundary conditions to simplify the calculations. Let us now consider other boundary conditions, starting with conditions on the lateral surface Σ.

3.2.1. *Boundary conditions on the lateral boundary* Σ

3.2.1.1. *Neumann boundary conditions*

Consider a Neumann boundary condition on the lateral boundary Σ: $\frac{\partial u}{\partial \nu}|_\Sigma = g$, where ν is the outward normal to Ω on Σ and g is in $L^2(0, a; H^{-1/2}(\partial\mathcal{O}))$. In this section, we will use the following spaces instead of the spaces in definition 1.2:

$$\tilde{X} = L^2(0,a;H^1(\mathcal{O})) \cap H^1(0,a;L^2(\mathcal{O})),$$

$$\tilde{X}_0 = \tilde{X} \cap \{u \in \tilde{X} : u|_{\Gamma_a} = 0\},$$

$$\tilde{Y} = \{u \in \tilde{X} : \frac{\partial^2 u}{\partial x^2} \in L^2(0,a;H^1(\mathcal{O})')\}.$$

By a result on the trace similar to proposition 1.1, $u|_{\Gamma_s}$ is defined in $H^{1/2}(\mathcal{O})$ and $\frac{\partial u}{\partial x}|_{\Gamma_s}$ is in $H^{-1/2}(\mathcal{O})$. We now consider the Dirichlet data h in $H^{1/2}(\mathcal{O})$, and the Neumann condition at $s = x$ is obtained in $H^{-1/2}(\mathcal{O})$. The right-hand side f is chosen from $L^2(0,a;H^1(\mathcal{O})')$.

Applying Green's formula [3.1] in a similar way to section 3.1 gives:

$$\int_0^x (\nabla_y u, \nabla_y \varphi)\, \mathrm{d}x' = \int_0^x < f, \varphi >_{H^1(\mathcal{O})' \times H^1(\mathcal{O})}\, \mathrm{d}x' + \int_\Sigma g\varphi\, \mathrm{d}\Sigma + \int_{\Gamma_x} \frac{\partial u}{\partial x}\varphi\, \mathrm{d}y + \int_{\Gamma_0} u_0 \varphi\, \mathrm{d}y, \qquad [3.8]$$

for all φ in $H^1(\mathcal{O})$. Consider the trajectory $u(.) \in L^2(0,a;H^1(\mathcal{O}))$. By an argument similar to that used in section 1.2, expression [3.4] now becomes the following, as the test function φ is linked to a Neumann boundary condition:

$$\left(\frac{\partial u}{\partial x}(x), \varphi\right) = (P(x)u(x),\varphi) + (w(x),\varphi)$$
$$\forall \varphi \in H^1(\mathcal{O}),\ \text{for a.e. } x \in]0,a[. \qquad [3.9]$$

By substituting [3.9] into [3.8] and evaluating as in section 3.1, differentiating w.r.t x and substituting in the expression for $\frac{\partial u}{\partial x}$, we find, for almost every $x \in [0,a]$:

$$\left(\frac{\mathrm{d}P}{\mathrm{d}x}u,\varphi\right) + (Pu, P\varphi) + (Pw,\varphi) + \left(\frac{\mathrm{d}w}{\mathrm{d}x},\varphi\right) =$$
$$= (\nabla_y u, \nabla_y \varphi) - \langle g, \varphi|_{\partial\mathcal{O}}\rangle_{H^{-1/2}(\partial\mathcal{O})\times H^{1/2}(\partial\mathcal{O})} - \langle f,\varphi\rangle_{H^1(\mathcal{O})'\times H^1(\mathcal{O})}. \qquad [3.10]$$

As the trajectory was chosen arbitrarily, we deduce as before the equation for P viewed as an operator in $\mathcal{L}(H^{1/2}(\mathcal{O}), H^{-1/2}(\mathcal{O}))$ and in $\mathcal{L}(H^1(\mathcal{O}), L^2(\mathcal{O}))$:

$$\left(\frac{\mathrm{d}P}{\mathrm{d}x}\varphi, \psi\right) + (P\varphi, P\psi) = (\nabla_y\varphi, \nabla_y\psi),$$
$$\forall\varphi, \psi \in H^1(\mathcal{O}), \quad P(0) = 0. \qquad [3.11]$$

The decoupled system [1.23–1.24] becomes

$$\left\langle \frac{\mathrm{d}w}{\mathrm{d}x} + Pw, \varphi \right\rangle_{H^1(\mathcal{O})' \times H^1(\mathcal{O})} = -\langle f, \varphi\rangle_{H^1(\mathcal{O})' \times H^1(\mathcal{O})}$$
$$-\langle g, \varphi|_{\partial\mathcal{O}}\rangle_{H^{-1/2}(\partial\mathcal{O}) \times H^{1/2}(\partial\mathcal{O})}, \quad \forall\varphi \in H^1(\mathcal{O}), w(0) = -u_0, \qquad [3.12]$$

$$\left(-\frac{\mathrm{d}u}{\mathrm{d}x}, \varphi\right) + (Pu, \varphi) = -(w, \varphi), \forall\varphi \in L^2(\mathcal{O}), u(a) = u_1. \qquad [3.13]$$

Note that we cannot interpret equations [3.12] and [3.13] in terms of boundary conditions on Σ. The condition satisfied by u on Σ is an inhomogeneous term that is taken into account by the equation of the residual w [3.12].

3.2.1.2. *Dirichlet boundary conditions*

We will first consider the case of a regular Dirichlet boundary condition on Σ. Consider the Dirichlet condition: $u|_\Sigma = u_b \in H^{1/2}(\Sigma)$. By definition of the space $H^{1/2}(\Sigma)$, there exists a lifting r of u_b such that:

$$r \in H^1(\Omega); \quad r|_\Sigma = u_b.$$

We will show that r may be chosen such that $\Delta r = 0$. Let $\tilde{r}$ be a lifting of u_b such that $\tilde{r} \in H^1(\Omega)$; $\tilde{r}|_\Sigma = u_b$. The Laplacian $\Delta\tilde{r}$ is in $H^{-1}(\Omega)$. Therefore, by solving the mixed problem:

$$-\Delta\overline{r} = \Delta\tilde{r}, \quad \overline{r}|_\Sigma = 0,$$

for example, with homogeneous Neumann boundary conditions on Γ_0 and Γ_a, we construct $r = \overline{r} + \tilde{r}$ (so that $\Delta r = 0$), which is as required.

Then, $\overline{u} = u - r$ satisfies:

$$\begin{cases} -\Delta \overline{u} = f \quad \text{in } \Omega, \\ \overline{u}|_\Sigma = 0, \quad -\dfrac{\partial \overline{u}}{\partial x}|_{\Gamma_0} = u_0 + \dfrac{\partial r}{\partial x}|_{\Gamma_0}, \quad \overline{u}|_{\Gamma_a} = u_1 - r|_{\Gamma_a}. \end{cases}$$

We now apply the previous method to $\overline{u}$. On each section of the cylinder, it holds that:

$$\frac{\partial \overline{u}}{\partial x} = P\overline{u} + \overline{w}.$$

By applying the method of section 1.2, we obtain the Riccati equation [1.22] for P:

$$\frac{\mathrm{d}P}{\mathrm{d}x} + P^2 + \Delta_y = 0, \quad P(0) = 0,$$

for the residual $\overline{w}$:

$$\frac{\mathrm{d}\overline{w}}{\mathrm{d}x} + P\overline{w} = -f, \quad \overline{w}(0) = -u_0 - \frac{\partial r}{\partial x}|_{\Gamma_0},$$

and for $\overline{u}$:

$$-\frac{\mathrm{d}\overline{u}}{\mathrm{d}x} + P\overline{u} = -\overline{w}, \quad \overline{u}(a) = u_1 - r|_{\Gamma_a}.$$

Finally, we have $u = \overline{u} + r$. The inhomogeneity of the Dirichlet condition on Σ is taken into account by the initial conditions of $\overline{w}$ and $\overline{u}$.

We will now consider a non-homogeneous Dirichlet boundary condition that is insufficiently regular for a lifting to exist: $u|_\Sigma = u_b \in L^2(\Sigma)$. We will go back over the principle of transposition and ultra-weak formulation of Lions Magenes [LIO 68c], but we will restrict attention to the spaces H^r defined in definition 2.1, with $H^0 = L^2(\mathcal{O}), H^1 = H_0^1(\mathcal{O})$. Starting from the equation satisfied by u, we choose a test function $\varphi \in H^2$ (independent of x), apply Green's formula [3.2] twice and integrate by parts w.r.t. x, which gives:

$$
\begin{aligned}
-\int_0^x (u, \Delta_y \varphi)_{H^0}\, \mathrm{d}x' = &\int_0^x < f, \varphi >_{H^{-1}(\mathcal{O}) \times H_0^1(\mathcal{O})}\, \mathrm{d}x' - \\
&-\int_0^x \left\langle u_b, \frac{\partial \varphi}{\partial \nu}|_{\partial \mathcal{O}} \right\rangle_{H^{-1/2}(\partial \mathcal{O}) \times H^{1/2}(\partial \mathcal{O})} \mathrm{d}x' + \\
&+\left\langle \frac{\partial u}{\partial x}(x), \varphi \right\rangle_{H^{-1} \times H^1} + \int_{\Gamma_0} u_0 \varphi\, \mathrm{d}y.
\end{aligned} \quad [3.14]
$$

By the same reasoning as above, we have:

$$
\left\langle \frac{\partial u}{\partial x}(x), \varphi \right\rangle_{H^{-1} \times H^1} = \langle P(x) u(x), \varphi \rangle_{H^{-1} \times H^1} + \langle w(x), \varphi \rangle_{H^{-1} \times H^1}
$$

$$
\forall \varphi \in H^1, \text{ for a.e. } x \in]0, a[, \quad [3.15]
$$

which we can substitute into [3.14]. By again evaluating the derivative w.r.t. x and substituting, after choosing the test function φ from H^2 (for simplicity, we omit dependency on x from the notation):

$$
\begin{aligned}
(u, -\Delta_y \varphi) - &< f, \varphi >_{H^{-1}(\mathcal{O}) \times H_0^1(\mathcal{O})} + \left\langle u_b, \frac{\partial \varphi}{\partial \nu}|_{\partial \mathcal{O}} \right\rangle_{H^{-1/2}(\partial \mathcal{O}) \times H^{1/2}(\partial \mathcal{O})} = \\
&= \left\langle \frac{\mathrm{d}P}{\mathrm{d}x} u, \varphi \right\rangle_{H^{-2} \times H^2} + \langle Pu, P\varphi \rangle_{H^{-1} \times H^1} + \\
&+ \langle Pw, \varphi \rangle_{H^{-2} \times H^2} + \left\langle \frac{\mathrm{d}w}{\mathrm{d}x}, \varphi \right\rangle_{H^{-2} \times H^2}.
\end{aligned} \quad [3.16]
$$

For P, viewed as an operator in $\mathcal{L}(H^1, H^0)$, in $\mathcal{L}(H^0, H^{-1})$ and in $\mathcal{L}(H^2, H^1)$, we obtain the following:

$$
\langle \frac{\mathrm{d}P}{\mathrm{d}x} \varphi, \psi \rangle_{H^{-2} \times H^2} + \langle P\varphi, P\psi \rangle_{H^{-1} \times H^1} = (\varphi, -\Delta_y \psi), \quad [3.17]
$$

$$
\forall \varphi \in H^0, \psi \in H^2, \quad P(0) = 0.
$$

The residual term w reflects the non-homogeneous assumptions on Σ:

$$\left\langle \frac{\mathrm{d}w}{\mathrm{d}x} + Pw, \varphi \right\rangle_{H^{-2}\times H^2} = - < f, \varphi >_{H^{-1}(\mathcal{O})\times H_0^1(\mathcal{O})}$$

$$+ \left\langle u_b, \frac{\partial \varphi}{\partial \nu}|_{\partial\mathcal{O}} \right\rangle_{H^{-1/2}(\partial\mathcal{O})\times H^{1/2}(\partial\mathcal{O})}, \quad \forall \varphi \in H^2,$$

$$w(0) = -u_0, \tag{3.18}$$

and u may be found from:

$$\langle -\frac{\mathrm{d}u}{\mathrm{d}x}, \varphi\rangle_{H^{-1}\times H^1} + \langle Pu, \varphi\rangle_{H^{-1}\times H^1} = -\langle w, \varphi\rangle_{H^{-1}\times H^1}, \tag{3.19}$$

$$\forall \varphi \in H^1, \quad u(a) = u_1.$$

3.2.1.3. *Robin boundary conditions*

In the case of a Robin boundary condition on Σ: $\frac{\partial u}{\partial \nu}|_\Sigma + \alpha u|_\Sigma = g$, equations [3.12] and [3.13] remain unchanged, but the Riccati equation [3.11] becomes:

$$\left(\frac{\mathrm{d}P}{\mathrm{d}x}\varphi, \psi\right) + (P\varphi, P\psi) = (\nabla_y \varphi, \nabla_y \psi) + \alpha\left(\varphi|_{\partial\mathcal{O}}, \psi|_{\partial\mathcal{O}}\right)_{L^2(\partial\mathcal{O})}$$

$$\forall \varphi, \psi \in H^1(\mathcal{O}), \quad P(0) = 0. \tag{3.20}$$

3.2.2. *Boundary conditions on the faces* Γ_0 *and* Γ_a

The condition on Γ_a used in the direct factorization [1.22–1.24] involving the Dirichlet-to-Neumann operator is the direct result of the Dirichlet boundary condition on this face in the initial problem $(\mathcal{P}_0)$. Neumann and Robin boundary conditions can be handled in the same way by using [1.17]. For example, for the Neumann condition:

$$\frac{\partial u}{\partial x}|_{\Gamma_a} = u_1 \tag{3.21}$$

the initial condition for backward integration w.r.t. u is given by:

$$P(a)u(a) = u_1 - w(a). \tag{3.22}$$

On the face Γ_0, we can handle Robin boundary conditions by direct invariant embedding. Indeed, for:

$$\left(-\frac{\partial u}{\partial x} + \alpha u\right)|_{\Gamma_0} = u_0,$$

the initial condition on P, which replaces [1.22], is given by:

$$P(0) = \alpha I.$$

But, for the residual w, the initial condition remains the same as [1.23]:

$$w(0) = -u_0.$$

However, it does not seem possible to process Dirichlet boundary conditions on Γ_0. Indeed, for the Dirichlet condition, we are led to write [1.24] as:

$$P(x)^{-1}\frac{\mathrm{d}u}{\mathrm{d}x}(x) = u(x) + P(x)^{-1}w(x), \qquad [3.23]$$

which non-rigorously leads to:

$$\lim_{x\to 0} P(x)^{-1} = 0. \qquad [3.24]$$

For backward invariant embedding as in section 1.3, we can also consider other types of boundary conditions on the faces Γ_0 and Γ_a. Difficulties only arise for Neumann boundary conditions on Γ_a for which there also exists a singularity in the initial condition on Q at a.

We will return to the question of singularities in the initial conditions of the Riccati equations in section 9.4, and we will discuss it numerically in remark 5.2, but note that this problem may also be approached with a Robin-to-Neumann operator as explained in section 3.2.3.

3.2.3. *Robin-to-Neumann operator*

We continue the discussion of the previous section regarding problem $(\mathcal{P}_0)$ with a Dirichlet condition on the boundary Γ_0:

$$(\mathcal{P}_0)\begin{cases} -\Delta u = -\dfrac{\partial^2 u}{\partial x^2} - \Delta_y u = f \quad \text{in } \Omega, \\ u|_\Sigma = 0, \quad u|_{\Gamma_0} = u_0, \quad u|_{\Gamma_a} = u_1. \end{cases} \qquad [3.25]$$

We embed this problem in the family $(\mathcal{P}_{s,h})$ with a Robin condition on the variable boundary Γ_s:

$$(\mathcal{P}_0)\begin{cases} -\Delta u = -\dfrac{\partial^2 u}{\partial x^2} - \Delta_y u = f \quad \text{in}\,\Omega, \\ u|_\Sigma = 0, \quad u|_{\Gamma_0} = u_0, \quad u|_{\Gamma_a} = u_1. \end{cases} \qquad [3.26]$$

Consider now the Robin-to-Neumann operator P' satisfying:

$$\frac{\partial u_{s,h}}{\partial x} = P'(s)h + w'(s). \qquad [3.27]$$

By a calculation similar to that in section 1.2, we obtain the Riccati equation for P':

$$\frac{\mathrm{d}P'}{\mathrm{d}x} + \beta P'^2 = \frac{P' - I}{\beta}\Delta_y(I - P'). \qquad [3.28]$$

Here, the advantage of this formulation is that we avoid the singularity at $x = 0$. Indeed, we have:

$$(I - P'(0))h = \beta u_0 + w'(0), \quad \forall h, \qquad [3.29]$$

from which we see that:

$$P'(0) = I, \quad w'(0) = -\beta u_0. \qquad [3.30]$$

The solution to [3.25] satisfies:

$$(I - P')\frac{\partial u}{\partial x} = \beta P' u + w', \qquad [3.31]$$

and the residual w' satisfies:

$$\frac{\partial w'}{\partial x} + \beta P' w' + \frac{1}{\beta}(I - P')\Delta_y(P'^{-1} - I)w' = -(I - P')f. \qquad [3.32]$$

3.2.4. *Neumann problem*

Consider problem $(\mathcal{P}_0)$ with the boundary conditions:

$$\frac{\partial u}{\partial \nu}|_{\Sigma} = g, \quad -\frac{\partial u}{\partial x}|_{\Gamma_0} = u_0, \quad \frac{\partial u}{\partial x}|_{\Gamma_a} = u_1.$$

To simplify the notation, we take $g \in L^2(\Sigma)$, $u_0, u_1 \in L^2(\mathcal{O})$ and $f \in L^2(\Omega)$. With these assumptions, given the lack of coercivity, we must place constraints on the data so that the problem is well-posed. The kernel of the adjoint operator is made up of constant functions. The Fredholm alternative implies that the data must satisfy:

$$\int_{\Omega} f \, \mathrm{d}x \, \mathrm{d}y + \int_{\mathcal{O}} u_0 \, \mathrm{d}y + \int_{\mathcal{O}} u_1 \, \mathrm{d}y + \int_{\Sigma} g \, \mathrm{d}\nu = 0. \qquad [3.33]$$

The solution of the problem is then determined up to an additive constant. Uniqueness is obtained by imposing an additional condition, such as:

$$\int_{\Gamma_0} u|_{\Gamma_0} \, \mathrm{d}y = 0. \qquad [3.34]$$

For the family of problems used to define invariant embedding, we can choose an arbitrary Neumann condition on the "moving" section Γ_s. We must then impose a compatibility condition on the Neumann data of the same type as [3.33] that varies as a function of s. It seems preferable to choose a Dirichlet condition that avoids the necessity of a compatibility condition, and so, for this problem, invariant embedding is performed in the same way as in section 3.2.1, yielding the factorized system [3.11–3.13]. As the problem that we wish to solve does not belong to the family of problems of the invariant embedding, we must find the initial condition at a for [3.13].

By the same argument based on the Fredholm alternative applied to the homogeneous problem analogous to [1.7] defined on Ω_s with a homogeneous Neumann condition on Σ, we show that P defined on $H^1(\mathcal{O})$ by [3.9] takes values in $\tilde{L}^2(\mathcal{O})$, where:

$$\tilde{L}^2(\mathcal{O}) = \{u \in L^2(\mathcal{O}) | \int_{\mathcal{O}} u \, \mathrm{d}y = 0\}.$$

It is possible to show that P is an isomorphism of $H^1(\mathcal{O})$ into this space. Then, w as defined by [3.9] satisfies compatibility equation [3.33] in the sense that:

$$\int_{\Omega} f \, dx \, dy + \int_{\mathcal{O}} u_0 \, dy + \int_{\mathcal{O}} w(a) \, dy + \int_{\Sigma} g \, d\nu = 0. \qquad [3.35]$$

In fact, it is enough to choose $\varphi = 1$ in [3.12], integrate this expression from 0 to a and use the fact that:

$$\int_{\mathcal{O}} Pw(x) \, dy = 0.$$

From [3.33] and [3.35], we deduce that:

$$\int_{\mathcal{O}} w(a) \, dy = \int_{\mathcal{O}} u_1 \, dy,$$

and so $w(a) - u_1$ belongs to $\tilde{L}^2(\mathcal{O})$. By writing out [3.13] on Γ_a, we obtain:

$$(-u_1, \varphi) + (P(a)u(a), \varphi) = -(w(a), \varphi), \quad \forall \varphi \in H^1. \qquad [3.36]$$

This allows the initial condition $u(a)$ for equation [3.13] to be uniquely determined as the solution in H^1 of:

$$P(a)u(a) = u_1 - w(a).$$

REMARK 3.1.– The process of invariant embedding yields a constraint analogous to [3.35] satisfied by $w(x)$ for all x:

$$\int_{\Omega_x} f \, dx \, dy + \int_{\mathcal{O}} u_0 \, dy + \int_{\mathcal{O}} w(x) \, dy + \int_{\Sigma_x} g \, d\nu = 0.$$

3.3. Explicitly taking into account the boundary conditions and the right-hand side

To simplify the notation in this section, we assume that $f \in L^2(\Omega)$. In the factorization [1.22–1.24], the residual term w represents both the effect of the condition on Γ_0, $-\frac{\partial u}{\partial x}|_{\Gamma_0} = u_0$ and the effect of the right-hand side

f. As was shown in remark 1.4, a change in u_1 affects only equation [1.24], which gives the evolution of u. In certain situations, it may be desirable to explicitly show the effect of this piece of data on the solution. We will begin by studying the case of a variation in u_0. We therefore define the operator $R(0, s) : H^{1/2}_{0,0}(\mathcal{O})' \to H^{1/2}_{0,0}(\mathcal{O})'$ as:

$$R(0, s)h = \frac{\partial \gamma'_{s,h}}{\partial x}|_{\Gamma_s},$$

where $\gamma'_{s,h}$ is a solution of:

$$\begin{cases} -\Delta \gamma'_{s,h} = 0 \quad \text{in } \Omega_s, \\ \gamma'_{s,h}|_{\Sigma_s} = 0, \quad -\dfrac{\partial \gamma'_{s,h}}{\partial x}|_{\Gamma_0} = h, \quad \gamma'_{s,h}|_{\Gamma_s} = 0, \end{cases} \tag{3.37}$$

which maps the Neumann condition h on Γ_0 to the normal derivative $\dfrac{\partial \gamma'_{s,h}}{\partial x}$ on Γ_x. The notation $R(0, x)$ indicates that the Neumann condition should be taken to be equal to u_0 at $x = 0$. In the next part of this section, we will link this operator to the evolution operator described by Pazy [PAZ 83] for the equation of w. However, let us first define the residual $w'(s)$ in $H^{1/2}_{0,0}(\mathcal{O})'$ as:

$$w'(s) = \frac{\partial \beta'_s}{\partial x}|_{\Gamma_s},$$

where β'_s is the solution of:

$$\begin{cases} -\Delta \beta'_s = f \quad \text{in } \Omega_s, \\ \beta'_s|_{\Sigma} = 0, \quad -\dfrac{\partial \beta'_s}{\partial x}|_{\Gamma_0} = 0, \quad \beta'_s|_{\Gamma_s} = 0. \end{cases} \tag{3.38}$$

The residual w' therefore depends only on f. The identity [1.17] becomes:

$$\frac{\mathrm{d}u}{\mathrm{d}x}(x) = P(x)u(x) + R(0, x)u_0 + w'(x). \tag{3.39}$$

We proceed by invariant embedding as in section 1.2, except that now the considered family of problems depends not just on the domain Ω_s and the Dirichlet data at Γ_s, but also on the Neumann value u_0 at Γ_0. We therefore consider the family of trajectories defined in section 1.2 except that the data u_0

is allowed to vary. Evaluating the derivative along a trajectory with fixed u_0, we have:

$$\frac{\mathrm{d}^2 u}{\mathrm{d}x^2} = -\Delta_y u - f = \frac{\mathrm{d}P}{\mathrm{d}x} u + P \frac{\mathrm{d}u}{\mathrm{d}x} + \frac{\mathrm{d}R}{\mathrm{d}x}(0,x) u_0 + \frac{\mathrm{d}w'}{\mathrm{d}x}.$$

Replacing $\frac{\mathrm{d}u}{\mathrm{d}x}$ with the expression in [3.39] and using the fact that $u(x)$ and u_0 were chosen arbitrarily, we arrive at the factorized form:

$$\begin{cases} \frac{\mathrm{d}P}{\mathrm{d}x} + P^2 + \Delta_y = 0, & P(0) = 0, \\ \frac{\mathrm{d}R}{\mathrm{d}x}(0,x) + P(x)R(0,x) = 0, & R(0,0) = -I, \\ \frac{\mathrm{d}w'}{\mathrm{d}x} + Pw' = -f, & w'(0) = 0, \\ -\frac{\mathrm{d}u}{\mathrm{d}x} + Pu = -w' - R(0,x)u_0, & u(a) = u_1. \end{cases} \qquad [3.40]$$

To interpret the operator $R(0,x) \in \mathcal{L}(H_{00}^{1/2}(\mathcal{O})', H_{00}^{1/2}(\mathcal{O})')$ appearing in the second equation of [3.40], we introduce Cauchy problem [1.23] with $f = 0$:

$$\frac{\mathrm{d}w}{\mathrm{d}x} + Pw = 0, \quad w(0) = -u_0. \qquad [3.41]$$

Then, $R(0,x)$ is the opposite of the evolution operator of w associated with equation [3.41]. Indeed, let:

$$w(x) = -R(0,x)w(0) = R(0,x)u_0.$$

The function w given by this formula is indeed the unique solution of [3.41]:

$$\frac{\mathrm{d}w}{\mathrm{d}x} = -\frac{\mathrm{d}R(0,x)}{\mathrm{d}x} u_0 = P(x)R(0,x)u_0 = -P(x)w(x), \quad w(0) = -u_0.$$

This formulation allows, for example, the "outgoing" flux at Γ_a to be directly calculated as a function of the "incoming" flux at Γ_0 using the operator $R(0,a)$:

$$\frac{\mathrm{d}u}{\mathrm{d}x}(a) = P(a)u_1 + w'(a) + R(0,a)u_0.$$

Consider now the full dependency of u on u_0, u_1 and f. As was defined just above, let $R(t,x)$ be the opposite of the evolution operator for a solution w of [3.41] on $]t,a[$ (we have $w(x) = -R(t,x)w(t)$) and let $S(s,x)$ be the evolution operator associated with the equation:

$$-\frac{\mathrm{d}\varphi}{\mathrm{d}x} + P\varphi = 0, \quad \varphi(s) \text{ given,} \qquad [3.42]$$

on $]0,s[$. Then, we have $\varphi(x) = S(s,x)\varphi(s)$. By the above argument, these operators satisfy:

$$\frac{\partial R}{\partial x}(t,x) + P(x)R(t,x) = 0, \quad t \leq x \leq a,\ R(t,t) = -I, \qquad [3.43]$$

$$-\frac{\partial S}{\partial x}(s,x) + P(x)S(s,x) = 0, \quad 0 \leq x \leq s,\ S(s,s) = I. \qquad [3.44]$$

Note that we can establish a link between $R(t,x)$ and $S(t,x)$. Differentiating [3.43] w.r.t. t gives:

$$\frac{\partial^2 R}{\partial x \partial t}(t,x) + P(x)\frac{\partial R}{\partial t}(t,x) = 0. \qquad [3.45]$$

It is clear that $\frac{\partial R}{\partial t}$, like R, is a solution of the differential equation [3.43] but with an initial condition at some other $x = t$. We can calculate this initial condition $\frac{\partial R}{\partial t}(t,t)$ as follows: by calculating the total derivative w.r.t. t of the initial condition $R(t,t) = I$, we obtain:

$$\frac{\partial R}{\partial t}(t,t) + \frac{\partial R}{\partial x}(t,t) = 0, \qquad [3.46]$$

and therefore:

$$\frac{\partial R}{\partial t}(t,t) = -\frac{\partial R}{\partial x}(t,t) = P(t)R(t,t) = -P(t). \qquad [3.47]$$

So, by linearity of the solution of [3.43] and the initial conditions satisfied by the solutions $R(t,x)$ and $\frac{\partial R}{\partial t}(t,x)$, we obtain the relation:

$$\frac{\partial R}{\partial t}(t,x) = P(t)R(t,x), \quad 0 \leq t \leq x \leq a, \quad R(t,t) = -I. \qquad [3.48]$$

By comparison of equations [3.44] and [3.48], and by uniqueness of the solutions of the differential equations, we deduce the relation:

$$R(t,x) = -S(x,t). \tag{3.49}$$

Let us return to the calculation of w'. The formula for the variation of the constant applied to the equation for w' in system [3.40] gives:

$$w'(x) = \int_0^x R(t,x) f(t)\,\mathrm{d}t. \tag{3.50}$$

Similarly, a solution u of [3.39] is given by:

$$u(x) = S(a,x)u_1 - \int_x^a S(t,x)(w'(t) + R(0,t)u_0)\,\mathrm{d}t. \tag{3.51}$$

So, from [3.50]:

$$\begin{aligned} u(x) = S(a,x)u_1 &- \int_x^a \int_0^t S(t,x)R(\tau,t)f(\tau)\,\mathrm{d}\tau\,\mathrm{d}t \\ &- \int_x^a S(t,x)R(0,t)u_0\,\mathrm{d}t. \end{aligned} \tag{3.52}$$

This formula explicitly produces the solution u as a function of u_0, u_1, f. We will consider this calculation again using the kernels of these operators in section 3.7.

We now wish to show that in the case of non-homogeneous Dirichlet boundary conditions on the boundary Σ not assumed to be regular (taken in $L^2(\Sigma)$), we can also explicitly show the direct contribution of these data in the solution. We will consider the case discussed in section 3.2.1.2. We consider the following modified version of [3.15]:

$$\begin{aligned} \left\langle \frac{\partial u}{\partial x}(x), \varphi \right\rangle_{H^{-1}\times H^1} &= \langle P(x)u(x), \varphi\rangle_{H^{-1}\times H^1} + \\ &\quad + \int_0^x \langle T(x',x)u_b(x'), \varphi\rangle_{H^{-1}\times H^1}\,\mathrm{d}x' \\ &\quad + \langle w'(x), \varphi\rangle_{H^{-1}\times H^1} \\ &\quad \forall \varphi \in H^1, \text{ i.e. for } x \in]0,a[, \end{aligned} \tag{3.53}$$

where $T(x', x)$ is an operator in $\mathcal{L}(L^2(\partial\mathcal{O}), H^{-1})$. Substituting [3.53] into [3.14] and taking the derivative w.r.t. x as in [3.16], we obtain:

$$\begin{aligned}
&(u, -\Delta_y \varphi) - < f, \varphi >_{H^{-1}(\mathcal{O}) \times H_0^1(\mathcal{O})} \\
&\quad + \left\langle u_b(x), \frac{\partial \varphi}{\partial \nu}|_{\partial\mathcal{O}} \right\rangle_{H^{-1/2}(\partial\mathcal{O}) \times H^{1/2}(\partial\mathcal{O})} \\
&= \left\langle \frac{\mathrm{d}P}{\mathrm{d}x} u, \varphi \right\rangle_{H^{-2} \times H^2} + \langle Pu, P\varphi \rangle_{H^{-1} \times H^1} \\
&\quad + \langle Pw', \varphi \rangle_{H^{-2} \times H^2} + \left\langle \frac{\mathrm{d}w'}{\mathrm{d}x}, \varphi \right\rangle_{H^{-2} \times H^2} \\
&\quad + \int_0^x < P(x) T(x', x) u_b(x'), \varphi >_{H^{-2} \times H^2} \mathrm{d}x' \\
&\quad + \int_0^x < \frac{\partial}{\partial x} T(x', x) u_b(x'), \varphi >_{H^{-2} \times H^2} \mathrm{d}x' \\
&\quad + < T(x, x) u_b(x), \varphi >_{H^{-1/2}(\partial\mathcal{O}) \times H^{1/2}(\partial\mathcal{O})} .
\end{aligned} \qquad [3.54]$$

By a formal identification argument that remains to be justified, using the fact that $u(x)$, $u_b(x')$ for $0 < x' < x$, $u_b(x)$ and φ are arbitrary, we obtain equations on P, T and w'. The operator P satisfies equation [3.17]. The operator $T(x', x)$ satisfies:

$$\begin{aligned}
&\frac{\mathrm{d}T}{\mathrm{d}x}(x', x) + P(x) T(x', x) = 0, \quad x' < x < a, \\
&\langle T(x, x)\psi, \varphi \rangle = \left(\psi, \frac{\partial \varphi}{\partial \nu}|_{\partial\mathcal{O}} \right), \quad \forall \psi \in L^2(\partial\mathcal{O}), \forall \varphi \in H^2.
\end{aligned} \qquad [3.55]$$

The residual w' satisfies:

$$\frac{\mathrm{d}w'}{\mathrm{d}x} + Pw' = -f, \quad w'(0) = -u_0, \qquad [3.56]$$

and u is given by solving:

$$\frac{\mathrm{d}u}{\mathrm{d}x} = Pu + \int_0^x T(x', x) ub(x') \, \mathrm{d}x' + w', \quad u(a) = u_1. \qquad [3.57]$$

If we compare these equations to the factorization [3.17–3.19] of this same problem, we observe that the effect of the Dirichlet condition u_b on Σ, which in [3.18] acted through the intermediary of the residual w, has now been made explicit by the action of the operator T.

3.4. Periodic boundary conditions in x

Consider now the following boundary value problem with boundary conditions that are periodic in x:

$$(\mathcal{P}_p)\begin{cases} -\Delta u = f \quad \text{in } \Omega, \\ u|_\Sigma = 0, \quad u|_{\Gamma_0} = u|_{\Gamma_a}, \quad \dfrac{\partial u}{\partial x}|_{\Gamma_0} = \dfrac{\partial u}{\partial x}|_{\Gamma_a}. \end{cases} \qquad [3.58]$$

We will show how R and S are useful in the factorization of this problem.

First of all, note that this problem is well-posed for $f \in L^2(\Omega)$. Let:

$$H^1_{0,p}(\Omega) = \{u \in H^1(\Omega) | u|_\Sigma = 0, u|_{\Gamma_0} = u|_{\Gamma_a}\}.$$

We can prove the Poincaré inequality in this space, from which it follows that the bilinear form associated with $(\mathcal{P}_p)$ is coercive. It now suffices to apply the Lax–Milgram lemma. We deduce that problem $(\mathcal{P}_p)$ has a unique solution in $H^1_{0,p}(\Omega)$.

Consider the operator from $(H^{1/2}_{00}(\mathcal{O}) \times (H^{1/2}_{00}(\mathcal{O})')$ to itself that sends $u(a)$ and $\dfrac{\mathrm{du}}{\mathrm{dx}}(0)$ to $u(0)$ and $\dfrac{\mathrm{du}}{\mathrm{dx}}(a)$, where u is the solution of problem $(\mathcal{P}_0)$ from chapter 1. This operator has a unique fixed point corresponding to the boundary conditions of the unique solution of problem $(\mathcal{P}_p)$.

Let us now perform invariant embedding as in section 3.3: suppose for now that $\dfrac{\mathrm{du}}{\mathrm{dx}}(0)$ is given and consider the family of boundary value problems on Ω_s with arbitrary Dirichlet data on Γ_s. As in section 3.3, we can separate the dependencies of the residual term on f and $\dfrac{\mathrm{du}}{\mathrm{dx}}(0)$, and hence impose $w'(0) = 0$. We then have:

$$\begin{cases} \dfrac{\mathrm{d}P}{\mathrm{d}x} + P^2 + \Delta_y = 0, & P(0) = 0, \\ \dfrac{\mathrm{d}R}{\mathrm{d}x}(0,x) + P(x)R(0,x) = 0, & R(0,0) = -I, \\ -\dfrac{\mathrm{d}S}{\mathrm{d}x}(a,x) + P(x)S(a,x) = 0, & S(a,a) = I, \\ \dfrac{\mathrm{d}w'}{\mathrm{d}x} + Pw' = -f, & w'(0) = 0, \\ -\dfrac{\mathrm{d}u}{\mathrm{d}x} + Pu = -w' + R(0,x)\dfrac{\mathrm{d}u}{\mathrm{d}x}(0). \end{cases} \qquad [3.59]$$

We must now ensure that for $s = a$, the considered problem indeed corresponds to $(\mathcal{P}_p)$. For this, we must find $\frac{\mathrm{du}}{\mathrm{dx}}(0)$ and $u(a)$ such that the periodicity conditions of $(\mathcal{P}_p)$ are satisfied. For the periodicity of the first derivative:

$$\frac{\mathrm{d}u}{\mathrm{d}x}(0) = -u_0 = \frac{\mathrm{d}u}{\mathrm{d}x}(a) = P(a)u(a) + w'(a) + R(0,a)u_0. \qquad [3.60]$$

For the periodicity of u, we use the formula given in [3.51]:

$$u(a) = u(0) = S(a,0)u(a) + \int_0^a S(t,0)(w'(t) + R(0,t)u_0)\,\mathrm{d}t. \quad [3.61]$$

Problem $(\mathcal{P}_p)$ may then be solved by decoupling as follows:

– we solve the fourth equation of [3.59] for w' in the direction of increasing x using the data f;

– we find u_0 and $u(a)$ by solving the system:

$$\begin{aligned} (I - S(a,0))u(a) - \left(\int_0^a S(t,0)R(0,t)\,\mathrm{d}t\right) u_0 &= \\ = \int_0^a S(t,0)w'(t)\,\mathrm{d}t&, \\ P(a)u(a) + (R(0,a) + I)u_0 = -w'(a)&. \end{aligned} \qquad [3.62]$$

The argument given above shows that this problem is well-posed;

– we solve the fifth equation of [3.59] with $u(a)$ in the direction of decreasing x, with $u(a)$ given and $\frac{\mathrm{d}u}{\mathrm{d}x}(0) = -u_0$.

3.5. An alternative but unstable formulation

In the invariant embedding calculation presented in section 1.2, we considered a family of boundary value problems $(\mathcal{P}_{s,h})$. We could also consider a family of Cauchy problems dependent on s and Cauchy data h and k on Γ_s:

$$(\mathcal{P'}_{s,h,k}) \begin{cases} -\Delta u_{s,h,k} = f \text{ in } \Omega_s, \\ u_{s,h,k}|_\Sigma = 0, \; u_{s,h,k}|_{\Gamma_s} = h, \; \dfrac{\partial u_{s,h,k}}{\partial x}|_{\Gamma_s} = k \in H_{00}^{1/2}(\mathcal{O})'. \end{cases} \quad [3.63]$$

It is a well-known fact that this problem is ill posed: first, the Cauchy data h and k cannot be independent and must satisfy a compatibility relation, and second, the solution does not depend continuously on the data, at least in the case of Sobolev space topologies. Indeed, from [3.39], h and k satisfy:

$$k - P(s)h - w'(s) \in ImR(0, s), \qquad [3.64]$$

where $\mathrm{Im}R(0, s)$ is the image of $H_{00}^{1/2}(\mathcal{O})'$ under $R(0, s)$. It can be shown that this image is dense in $H_{00}^{1/2}(\mathcal{O})'$ but is not a Sobolev-type space. It is nevertheless possible to perform invariant embedding for the family of Cauchy problems [3.63]. If we follow a trajectory as defined above, i.e. a function $x \longmapsto u(x) \in H_{00}^{1/2}(\mathcal{O})$ satisfying Poisson's equation and the condition $-\dfrac{\partial u}{\partial x}|_{\Gamma_0} = u_0$, the compatibility relation between the data h and k follows from relation [1.16]:

$$k = P(s)h + w(s). \qquad [3.65]$$

Indeed, if h and k are related via [3.65], the solution of $(\mathcal{P}'_{s,h,k})$ is also a solution of $(\mathcal{P}_{s,h})$. This invariant embedding leads us to integrate the equations for P and w [1.22–1.23] which allows us to calculate the Cauchy conditions at Γ_a. We thus obtain the alternative factorization:

$$\begin{cases} \dfrac{\mathrm{d}P}{\mathrm{d}x} + P^2 + \Delta_y = 0, & P(0) = 0, \\ \dfrac{\mathrm{d}w}{\mathrm{d}x} + Pw = -f, & w(0) = -u_0, \\ -\dfrac{\mathrm{d}^2u}{\mathrm{d}x^2} - \Delta_y u = f, & u(a) = u_1, \quad \dfrac{\mathrm{d}u}{\mathrm{d}x}(a) = P(a)u_1 + w(a). \end{cases} \qquad [3.66]$$

The Cauchy problem thus formulated in the third line of [3.66] admits a solution but is unstable. Note that although it is unstable, it has the advantage that the values of P do not need to be saved for the "upward" phase. Indeed, the only purpose of calculating P and w is to obtain $P(a)$ and $w(a)$ to fulfill the Cauchy conditions for u on Γ_a. We will see in Chapter 5, remark 5.4, that an analogous calculation for the discretized problem may be interpreted as a block Gaussian factorization with row permutations at each step. In other words, the continuous analog of row permutation in the Gaussian factorization is, in this case, the act of replacing [1.24] with the Cauchy problem in the third equation [3.66]. This establishes a link between the instability of the Cauchy problem for the Laplacian and the instability of the Gaussian factorization without pivoting in the corresponding discretized problem.

3.6. Link with the Steklov–Poincaré operator

In domain decomposition methods, we fairly often require the Steklov–Poincaré operator, which connects the boundary conditions on two adjacent subdomains (see the book by Quarteroni and Valli [QUA 99]). More precisely, for problem $(\mathcal{P}_0)$, consider the partition of the domain Ω into Ω_s and $\widetilde{\Omega}_s$ separated by Γ_s, as in sections 1.2 and 1.3. Let $h \in H_{00}^{1/2}(\mathcal{O})$ and $\gamma_{s,h}$ be given according to definition 1.2. Consider also $\tilde{\gamma}'_{s,h}$ satisfying definition 1.3, except that the Neumann condition on Γ_s is replaced by a Dirichlet condition:

$$\begin{cases} -\Delta\tilde{\gamma}'_{s,h} = 0 \quad \text{in } \widetilde{\Omega}_s, \\ \tilde{\gamma}'_{s,h}|_\Sigma = 0, \quad \tilde{\gamma}'_{s,h}|_{\Gamma_s} = h, \quad \tilde{\gamma}'_{s,h}|_{\Gamma_a} = 0. \end{cases}$$

We deduce that:

$$-\frac{\partial\tilde{\gamma}'_{s,h}}{\partial x}|_{\Gamma_s} = Q(s)^{-1}h,$$

where $Q(s)$ is the same as in definition 1.3 and satisfies equation [1.40]. We similarly define $\tilde{\beta}'_s$ by:

$$\begin{cases} -\Delta\tilde{\beta}'_s = f \quad \text{in } \widetilde{\Omega}_s, \\ \tilde{\beta}'_s|_\Sigma = 0, \quad \tilde{\beta}'_s|_{\Gamma_s} = 0, \quad \tilde{\beta}'_s|_{\Gamma_a} = u_1. \end{cases}$$

We consider the function $r(s) = \tilde{\beta}_s|_{\Gamma_s}$ defined in definition 1.3 and satisfying equation [1.41]. Then, for $\tilde{u} = \tilde{u}_{s,h}$, a solution of $(\tilde{\mathcal{P}}_{s,Q(s)^{-1}h})$, observing that $\tilde{\gamma}'_s = \tilde{\gamma}'_{s,h}$, on Γ_s, we have that:

$$\frac{\partial\tilde{u}}{\partial x}|_{\Gamma_s} = \frac{\partial\tilde{\beta}'_s}{\partial x}|_{\Gamma_s} + \frac{\partial\tilde{\gamma}'_s}{\partial x}|_{\Gamma_s}.$$

Thus, the Steklov–Poincaré operator $\mathcal{S}(s)$ is defined in $\mathcal{L}(H_{00}^{1/2}(\mathcal{O}), H_{00}^{1/2}(\mathcal{O})')$ by:

$$\mathcal{S}(s)h = \frac{\partial \gamma_s}{\partial x}|_{\Gamma_s} - \frac{\partial \tilde{\gamma}'_s}{\partial x}|_{\Gamma_s}, \qquad [3.67]$$

and the Steklov–Poincaré interface equation is:

$$\mathcal{S}(s)h = \frac{\partial \tilde{\beta}'_s}{\partial x}|_{\Gamma_s} - \frac{\partial \beta_s}{\partial x}|_{\Gamma_s}. \qquad [3.68]$$

By definitions 1.2 and 1.3, we have:

$$\frac{\partial \gamma_s}{\partial x}|_{\Gamma_s} = P(s)h, \quad -\frac{\partial \tilde{\gamma}'_s}{\partial x}|_{\Gamma_s} = Q(s)^{-1}h \qquad [3.69]$$

and:

$$\frac{\partial \beta_s}{\partial x}|_{\Gamma_s} = w(s), \quad \frac{\partial \tilde{\beta}'_s}{\partial x}|_{\Gamma_s} = Q(s)^{-1}r(s). \qquad [3.70]$$

It therefore holds that:

$$\mathcal{S}(s) = P(s) + Q(s)^{-1}, \qquad [3.71]$$

and after multiplying by $Q(s)$ for $s \neq a$, the Steklov–Poincaré equation becomes:

$$(Q(s)P(s) + I)h = r(s) - Q(s)w(s). \qquad [3.72]$$

REMARK 3.2.– In the context of domain decomposition methods, it can be desirable to modify the decomposition and move the boundary between subdomains in order to balance calculations. The calculations given in this section can help to calculate the change in the Steklov–Poincaré equation induced by this displacement. Let $T(s) = Q(s)P(s) + I$ be the operator appearing in the Steklov–Poincaré equation. From [1.22] and [1.40], we see that $T(s)$ satisfies:

$$\frac{\mathrm{d}T(s)}{\mathrm{d}s} = -P(s)T(s) - T(s)\Delta_y Q(s). \qquad [3.73]$$

If the values of P have previously been saved, for example, by substituting $Q(s)$ with $(T(s)-I)P(s)^{-1}$, this calculation allows the change in the Steklov–Poincaré equation when the interface Γ_s is moved to be obtained by integrating the Riccati equation from the point at which T is known:

$$\frac{\mathrm{d}T(s)}{\mathrm{d}s} = -P(s)T(s) + T(s)\Delta_y P(s)^{-1} - T(s)\Delta_y T(s)P(s)^{-1}.$$

3.7. Application of the Schwarz kernel theorem: link with Green's functions and Hadamard's formula

We return to the situation described in section 1.2. As in [LIO 68a], we can apply the Schwartz kernel theorem [SCH 59]. Let $P(x, y, \eta)$ be the kernel associated with $P(x)$, which is the distribution on $\mathcal{O}_y \times \mathcal{O}_\eta$ uniquely defined by:

$$(P(x)\varphi)(y) = \int_{\mathcal{O}} P(x, y, \eta)\varphi(\eta)\,\mathrm{d}\eta, \quad \forall\varphi \in \mathcal{D}(\mathcal{O}). \quad [3.74]$$

PROPOSITION 3.1.– The Riccati equation [1.22] on the operator $P(x)$ becomes an equation on the kernel:

$$\frac{\partial P(x, y, \eta)}{\partial x} + \int_{\mathcal{O}} P(x, y, \zeta)P(x, \zeta, \eta)\,\mathrm{d}y + \Delta_y \delta(y - \eta) = 0,$$

$$P(0, y, \eta) = 0, \quad [3.75]$$

where $\delta(y - \eta)$ is the kernel of the identity. By self-adjointness of $P(x)$, we also have:

$$P(x, y, \eta) = P(x, \eta, y), \quad [3.76]$$

and from the boundary conditions on the lateral boundary Σ:

$$P(x, y, \eta) = 0, \quad y \in \Sigma, \quad \eta \in \Omega, \quad P(x, y, \eta) = 0, \quad y \in \Omega, \quad \eta \in \Sigma. \quad [3.77]$$

The equations on w and u [1.23–1.24] may be written as:

$$\frac{\partial w(x, y)}{\partial x} + \int_{\mathcal{O}} P(x, y, \eta)w(x, \eta)\,\mathrm{d}\eta = -f(x, y),\ w(0, y) = -u_0(y), \quad [3.78]$$

$$-\frac{\partial u(x, y)}{\partial x} + \int_{\mathcal{O}} P(x, y, \eta)u(x, \eta)\,\mathrm{d}\eta = -w(x, y), \quad u(a, y) = u_1(y). \quad [3.79]$$

PROOF.– We will only verify the second term of [3.75]. The kernel associated with $P(x)^2$ is defined by:

$$\begin{aligned}(P(x)^2\varphi)(y) &= \int_{\mathcal{O}} P(x,y,\eta)(P(x)\varphi)(\eta)\,\mathrm{d}\eta \\ &= \int_{\mathcal{O}} P(x,y,\eta)\left(\int_{\mathcal{O}} P(x,\eta,\zeta)\varphi(\zeta)\,\mathrm{d}\zeta\right)\mathrm{d}\eta \\ &= \int_{\mathcal{O}}\left(\int_{\mathcal{O}} P(x,y,\eta)P(x,\eta,\zeta)\,\mathrm{d}\eta\right)\varphi(\zeta)\,\mathrm{d}\zeta.\end{aligned}$$

This kernel is therefore given by $\int_{\mathcal{O}} P(x,y,\eta)P(x,\eta,\zeta)\,\mathrm{d}\eta$. □

Returning to the calculations in section 3.3, we introduce the kernels $R(t,x,y,\eta)$ and $S(s,x,y,\eta)$ of $R(t,x)$ and $S(s,x)$, respectively, defined by:

$$R(t,x)\varphi = \int_{\mathcal{O}} R(t,x,y,\eta)\varphi(\eta)\,\mathrm{d}\eta, \quad S(s,x)\varphi = \int_{\mathcal{O}} S(s,x,y,\eta)\varphi(\eta)\,\mathrm{d}\eta, \quad \forall\varphi \in \mathcal{D}(\mathcal{O}). \qquad [3.80]$$

Equations [3.43–3.44] become:

$$\frac{\mathrm{d}R}{\mathrm{d}x}(t,x,y,\eta) + \int_{\mathcal{O}} P(x,y,y')R(t,x,y',\eta)\,\mathrm{d}y' = 0, \qquad [3.81]$$

$$R(t,t,y,\eta) = -\delta(y-\eta),$$

$$-\frac{\mathrm{d}S}{\mathrm{d}x}(s,x,y,\eta) + \int_{\mathcal{O}} P(x,y,y')S(s,x,y',\eta)\,\mathrm{d}y' = 0, \qquad [3.82]$$

$$S(s,s,y,\eta) = \delta(y-\eta).$$

Formula [3.52] may be rewritten as:

$$\begin{aligned}u(x,y) &= \int_{\mathcal{O}} S(a,x,y,\eta)u_1(\eta)\,\mathrm{d}\eta \\ &\quad - \int_{\mathcal{O}}\int_{\mathcal{O}}\int_x^a\int_0^t S(t,x,y,y')R(\tau,t,y',\eta)f(\tau,\eta)\,\mathrm{d}\tau\,\mathrm{d}t\,\mathrm{d}\eta\,\mathrm{d}y' \\ &\quad - \int_{\mathcal{O}}\int_{\mathcal{O}}\int_x^a S(t,x,y,y')R(0,t,y',\eta)u_0(\eta)\,\mathrm{d}t\,\mathrm{d}\eta\,\mathrm{d}y'. \qquad [3.83]\end{aligned}$$

By changing the order of integration:

$$u(x,y) = \int_{\mathcal{O}} S(a,x,y,\eta)u_1(\eta)\,\mathrm{d}\eta$$
$$- \int_{\mathcal{O}} \int_0^a \int_{\mathcal{O}} \int_{x\wedge\tau}^a S(t,x,y,y')R(\tau,t,y',\eta)f(\tau,\eta)\,\mathrm{d}t\,\mathrm{d}\eta\,\mathrm{d}\tau\,\mathrm{d}y'$$
$$- \int_{\mathcal{O}} \int_{\mathcal{O}} \int_x^a S(t,x,y,y')R(0,t,y',\eta)u_0(\eta)\,\mathrm{d}t\,\mathrm{d}\eta\,\mathrm{d}y',$$

where $x \wedge \tau$ denotes $\max(x,\tau)$. Consider now Green's kernel $G(a;x,y;\xi,\eta)$ associated with problem $(\mathcal{P}_0)$ on Ω. Green's representation formula may be written as:

$$u(x,y) = \int_{\Omega} G(a;x,y;\xi,\eta)f(\xi,\eta)\,\mathrm{d}\xi\,\mathrm{d}\eta + \int_{\Gamma_0} G(a;x,y;0,\eta)u_0(\eta)\,\mathrm{d}\eta$$
$$- \int_{\Gamma_a} \frac{\partial G}{\partial \xi}(a;x,y;\xi,\eta)|_{\xi=a} h(\eta)\,\mathrm{d}\eta. \qquad [3.84]$$

By comparing between these two formulas (u_0, u_1, f are arbitrary), we deduce that:

$$G(a;x,y;\xi,\eta) = -\int_{\mathcal{O}} \int_x^a S(t,x,y,y')R(\xi,t,y',\eta)\,\mathrm{d}t\,\mathrm{d}y'. \qquad [3.85]$$

Indeed, we see that:

$$\frac{\partial G}{\partial \xi}(a;x,y;\xi,\eta) = -\int_{\mathcal{O}} \int_x^a S(t,x,y,y')\frac{\partial R}{\partial \xi}(\xi,t,y',\eta)\,\mathrm{d}t\,\mathrm{d}y$$
$$+ \begin{cases} \int_{\mathcal{O}} S(\xi,x,y,y')R(\xi,\xi,y',\eta)\,\mathrm{d}y' & \text{if } \xi > x \\ 0 & \text{if } \xi < x. \end{cases}$$

And hence:

$$\frac{\partial G}{\partial \xi}(a;x,y;\xi,\eta)|_{\xi=a} = -S(a,x,y,\eta). \qquad [3.86]$$

REMARK 3.3.– As will be explored in more detail in Chapter 5, if formulation [3.75], [3.78–3.79] can be viewed as an infinite-dimensional extension of the block LU factorization of a block tridiagonal matrix, the formula in [3.83] corresponds to the generalization of the expression for the inverse of these matrices. It is a well-known fact in numerical analysis that calculating the inverse is more expensive than LU factorization. Indeed, if we assume that the factorization is known, calculating the inverse is equivalent to the solution of n linear systems from the factorized forms of the n basis vectors. The same remark holds here, as in order to obtain formula [3.83], $R(x, \xi, y, \eta)$ and $S(x, \xi, y, \eta)$ must be calculated for each couple x, ξ.

Consider now the derivative of the Green's function w.r.t. the domain size:

$$\frac{\partial G}{\partial a}(a; x, y; \xi, \eta) = -\int_{\mathcal{O}} S(a, x, y, y')R(\xi, a, y', \eta)\,\mathrm{d}y'. \qquad [3.87]$$

Now, we can show as in [3.86] that:

$$\frac{\partial G}{\partial x}(a; x, y; \xi, \eta)|_{x=a} = R(\xi, a, y, \eta). \qquad [3.88]$$

We then deduce from [3.86–3.88] that:

$$\frac{\partial G}{\partial a}(a; x, y; \xi, \eta) = \int_{\mathcal{O}} \frac{\partial G}{\partial \xi}(a; x, y; \xi, y')|_{\xi=a}\frac{\partial G}{\partial x}(a; x, y'; \xi, \eta)|_{x=a}\,\mathrm{d}y', \qquad [3.89]$$

which, for the case of the cylinder considered here, is exactly Hadamard's formula for the derivative of Green's function w.r.t. the variation of the domain.

REMARK 3.4.– Together with the values of R and S calculated using [3.75], [3.81–3.82], the formula in [3.87] allows $G(a; x, y; \xi, \eta)$ to be calculated for a ranging from 0 to the desired value. This method, excluding the calculation of R and S, was suggested in Hadamard's original paper [HAD 10] as a practical method of calculating Green's function.

Similarly, for the backward invariant embedding calculation in section 1.3, we can define the kernel $Q(x, y, \eta)$ of the Neumann-to-Dirichlet operator Q

by:

$$(Q(x)\varphi)(y) = \int_{\mathcal{O}} Q(x,y,\eta)\varphi(\eta)\,\mathrm{d}\eta, \quad \forall\varphi \in \mathcal{D}(\mathcal{O}). \qquad [3.90]$$

In this context, the equations for Q, r and u [1.40–1.42] become:

$$\frac{\partial Q(x,y,\eta)}{\partial x} + \int_{\mathcal{O}} Q(x,y,\zeta)\Delta_\zeta Q(x,\zeta,y)\,\mathrm{d}\zeta + \delta(y-\eta) = 0,$$

$$Q(a,y,\eta) = 0,$$

$$\frac{\partial r(x,y)}{\partial x} + \int_{\mathcal{O}} Q(x,y,\eta)\Delta_\eta r(x,\eta)\,\mathrm{d}\eta = -\int_{\mathcal{O}} Q(x,y,\eta) f(x,\eta)\,\mathrm{d}\eta,$$

$$r(a,y) = u_1(y),$$

$$\int_{\mathcal{O}} Q(x,y,\eta)\frac{\partial u(x,\eta)}{\partial x}\,\mathrm{d}\eta + u(x,y) = r(x,y),$$

$$u(0,y) = \int_{\mathcal{O}} Q(0,y,\eta)u_0(\eta)\,\mathrm{d}\eta + r(0,y).$$

4

Interpretation of the Factorization through a Control Problem

In section 1.2, we presented a factorization of problem $(\mathcal{P}_0)$ in a cylinder involving a Riccati equation. This equation is reminiscent of the equation that yields the optimal feedback for a control problem with linear dynamics and quadratic cost. In Chapter 2, we also studied this equation by adapting the method developed by J.-L. Lions in [LIO 68a] to solve closed-loop optimal control problems in systems determined by parabolic equations. In this chapter, we will give an equivalent formulation of problem $(\mathcal{P}_0)$ in terms of an optimal control problem. We will show that decoupling this control problem produces the same decoupled system as in section 1.2. This approach provides a different kind of insight into the analytical difficulties posed by the Riccati equation [1.22] due to the unboundedness of the operator P. However, note that this approach of reformulating the problem in terms of optimal control is limited to self-adjoint boundary value problems.

4.1. Formulation of problem $(\mathcal{P}_0)$ in terms of optimal control

We formulate a dynamic problem, with x, the coordinate along the axis of the cylinder Ω, playing the role of a time variable. The state equation is given by a simple integration operator in x. To maintain consistency with the notation of section 1.1, we consider a problem in the backward direction with the final state given at $x = a$. Consider the control space $\mathcal{U} = L^2(\Omega)$. For $v \in \mathcal{U}$, the state $u(v)$ satisfies:

$$\begin{cases} -\dfrac{\partial u}{\partial x} = v & \text{in } \Omega, \\ u(a) = u_1. \end{cases} \qquad [4.1]$$

The solution of this state equation is given by:

$$u(x) = u_1 + \int_x^a v(z)\, \mathrm{d}z. \qquad [4.2]$$

We denote a function g defined on Ω by $g(x) = g(x, .)$ to emphasize the dependency on x. We also write $u(v)$ the state u associated with the control v via [4.1]. The specific circumstances of each situation easily lead us to the correct definition in the following.

For all $v \in \mathcal{U}$, the state $u(v)$ is an element of $H^1(0, a; L^2(\mathcal{O}))$. Note that with this definition, the state does not have $H^1(\mathcal{O})$ regularity in the transversal variable y and, in particular, we cannot impose a Dirichlet condition $u|_\Sigma = 0$ on the lateral boundary. Thus, we will proceed as in [LIO 83] by reducing the space of admissible controls so that the solution has the desired regularity. We therefore define the set of admissible controls by $\mathcal{U}_{ad} = \{v \in \mathcal{U} : u(v) \in X_{u_1}\}$, where:

$$X_{u_1} = \{h \in L^2(0, a; H_0^1(\mathcal{O})) \cap H^1(0, a; L^2(\mathcal{O})) : h(a) = u_1\}.$$

Note that $\mathcal{U}_{ad}$ is dense in $\mathcal{U}$. The desired state u_d is given for almost every x by the solution of the family of Dirichlet problems on the cylindrical sections of dimension $(n - 1)$, with the assumption of Chapter 1 that $f \in \hat{X}' = L^2(0, a; H^{-1}(\mathcal{O}))$:

$$\begin{cases} -\Delta_y u_d(x) = f(x) & \text{in } \mathcal{O} \\ u_d|_{\partial\mathcal{O}} = 0. \end{cases} \qquad [4.3]$$

Then, u_d belongs to $L^2(0, a; H_0^1(\mathcal{O}))$. We define the cost function $J(v)$ of the control problem by:

$$\begin{aligned} J(v) = &\int_0^a \|\nabla_y u(v) - \nabla_y u_d\|^2_{L^2(\mathcal{O})}\, \mathrm{d}x + \int_0^a\!\!\int_{\mathcal{O}} v^2\, \mathrm{d}x\, \mathrm{d}y \\ &- 2 < u_0, u(v)|_{\Gamma_0} >_{H_{0,0}^{1/2}(\mathcal{O})' \times H_{0,0}^{1/2}(\mathcal{O})}. \end{aligned} \qquad [4.4]$$

We now assume that $u_0 \in L^2(\mathcal{O})$. For the case where $u_0 \in (H^{1/2}_{0,0}(\mathcal{O}))'$, we refer to remark 4.2.

For $v \in \mathcal{U}_{ad}$, the functional J is well defined. The optimal control problem consists of finding $v^\star \in \mathcal{U}_{ad}$, such that:

$$J(v\star) = \min_{v \in \mathcal{U}ad} J(v), \qquad [4.5]$$

and we write $u^\star = u(v^\star)$. As $\mathcal{U}_{ad}$ is not a closed subspace of $L^2(\Omega)$, we cannot apply classical results that guarantee the existence of a solution to this problem (such as those used by [LIO 68a]).

PROPOSITION 4.1.– Problem [4.5] admits a unique solution $v^\star \in \mathcal{U}_{ad}$.

PROOF.– The functional $J(v)$ may be rewritten as a functional of the state u by applying state equation [4.1]. It holds that:

$$J(v) = \overline{J}(u) = \int_0^a \|\nabla_y u - \nabla_y u_d\|^2_{L^2(\mathcal{O})} \, \mathrm{d}x + \int_0^a \int_{\mathcal{O}} \left| \frac{\partial u}{\partial x} \right|^2 \, \mathrm{d}x \, \mathrm{d}y$$
$$- 2 \int_{\Gamma_0} u_0 u(0, y; v) \, \mathrm{d}y. \qquad [4.6]$$

But now, $\overline{J}(u)$ is a strictly convex continuous quadratic functional on the Hilbert space $L^2(0, a; H^1_0(\mathcal{O})) \cap H^1(0, a; L^2(\mathcal{O}))$. The affine (and therefore convex) subspace X_{u_1} is closed in $L^2(0, a; H^1_0(\mathcal{O})) \cap H^1(0, a; L^2(\mathcal{O}))$. Hence, there exists a unique $u^\star$ such that:

$$\overline{J}(u^\star) = \min_{h \in X_{u_1}} \overline{J}(h).$$

Let $v^\star = \dfrac{\partial u^\star}{\partial x}$. Then, $v^\star$ belongs to $\mathcal{U}_{ad}$ and

$$J(v^\star) = \overline{J}(u^\star) = \min_{h \in X_{u_1}} \overline{J}(h) = \min_{v \in \mathcal{U}_{ad}} J(v). \qquad \square$$

PROPOSITION 4.2.–The state $u^\star$, constructed from state equation [4.1] and the optimal control $v^\star$, is characterized by:

$$\int_0^a < -\Delta_y u^\star - f, h >_{H^{-1}(\mathcal{O}) \times H^1_0(\mathcal{O})} \, \mathrm{d}x + \int_0^a \int_{\mathcal{O}} \frac{\partial u^\star}{\partial x} \frac{\partial h}{\partial x} \, \mathrm{d}x \, \mathrm{d}y$$
$$- \int_{\Gamma_0} u_0 h \, \mathrm{d}y = 0, \quad \forall h \in X_0. \qquad [4.7]$$

Furthermore, the state $u^\star$ is a solution of problem $(\mathcal{P}_0)$.

PROOF.– The functional $\overline{J}$ is strictly convex and Fréchet differentiable. Let $\overline{J}'$ be its differential. The unique minimum $u^\star$ is characterized by:

$$\overline{J}'(u^\star)(h) = 2\int_0^a\int_{\mathcal{O}}(\nabla_y u^\star - \nabla_y u_d).\nabla_y h \,\mathrm{d}x\,\mathrm{d}y$$
$$+\, 2\int_0^a\int_{\mathcal{O}} \frac{\partial u^\star}{\partial x}\frac{\partial h}{\partial x}\,\mathrm{d}x\,\mathrm{d}y - 2\int_{\Gamma_0} u_0 h\,\mathrm{d}y = 0 \quad \forall\, h \in X_0. \qquad [4.8]$$

From definition [4.3] of u_d, we obtain relation [4.7]. To check that the state $u^\star$ is a solution of problem $(\mathcal{P}_0)$, it is sufficient to observe that [4.8] corresponds to its variational formulation. Equivalently, we can achieve the same result by arguing on the functional J directly: from relation [4.6], develop the square of the first term on the right-hand side. Regrouping with the second term:

$$\overline{J}(u) = \int_\Omega |\nabla u|^2 \;\mathrm{d}x\;\mathrm{d}y - 2\int_0^a (\nabla_y u, \nabla_y u_d)_{L^2(\mathcal{O})}\;\mathrm{d}x$$
$$+ \int_0^a \|\nabla_y u_d\|^2_{L^2(\mathcal{O})}\;\mathrm{d}x - 2\int_{\Gamma_0} u_0(y)u(0,y;v)\;\mathrm{d}y.$$

Once again, from definition [4.3] of u_d, we have that:

$$\overline{J}(u) = \int_\Omega |\nabla u|^2 \;\mathrm{d}x\;\mathrm{d}y - 2\int_\Omega fu\;\mathrm{d}x\;\mathrm{d}y + \int_0^a \|\nabla_y u_d\|^2_{L^2(\mathcal{O})}\;\mathrm{d}x$$
$$- 2\int_{\Gamma_0} u_0 u(0,y;v)\;\mathrm{d}y.$$

It is now clear that, up to a constant term, $\overline{J}(u)$ is the energy functional of problem $(\mathcal{P}_0)$. □

We introduce the adjoint state p of the control problem by:

$$\begin{cases} \dfrac{\partial p}{\partial x} = -\Delta_y u - f & \text{in } \Omega, \\ p(0) = -u_0. \end{cases} \qquad [4.9]$$

Then, since $-\Delta_y u - f \in L^2(0,a;H^{-1}(\mathcal{O}))$, we know that p belongs to $H^1(0,a;H^{-1}(\mathcal{O}))$ (see theorem 1.2 from Chapter 3 of [LIO 68a]). Furthermore, since u is a member of the space Y defined in [1.2], we can also deduce that:

$$\frac{\partial p}{\partial x} = \frac{\partial^2 u}{\partial x^2} \in H^{-1}(0,a;L^2(\mathcal{O})),$$

and hence that $p \in L^2(\Omega)$.

Consider the adjoint state $p^\star$ associated with the optimal state $u^\star$. From the optimality condition [4.7], we deduce that:

$$\int_0^a < -\Delta_y u^\star - f, h >_{H^{-1}(\mathcal{O})\times H_0^1(\mathcal{O})} \,\mathrm{d}x + \int_0^a\int_{\mathcal{O}} \frac{\partial u^\star}{\partial x}\frac{\partial h}{\partial x}\,\mathrm{d}x\,\mathrm{d}y - \int_{\Gamma_0} u_0 h\,\mathrm{d}y$$
$$= \int_0^a < \frac{\partial p^\star}{\partial x}, h >_{H^{-1}(\mathcal{O})\times H_0^1(\mathcal{O})} \,\mathrm{d}x + \int_0^a\int_{\mathcal{O}} \frac{\partial u^\star}{\partial x}\frac{\partial h}{\partial x}\,\mathrm{d}x\,\mathrm{d}y - \int_{\Gamma_0} u_0 h\,\mathrm{d}y = 0.$$

Integrating by parts w.r.t. x:

$$\int_0^a\int_{\mathcal{O}} (-p^\star + \frac{\partial u^\star}{\partial x})\, v\,\mathrm{d}x\,\mathrm{d}y = 0, \quad \forall\, v = \frac{\partial h}{\partial x} \in \mathcal{U}ad, \text{where } h(a) = 0. \quad [4.10]$$

Hence, the optimal control is given as a function of the adjoint state by:

$$v^\star = -p^\star. \quad [4.11]$$

Thus, we have obtained the optimality system of the control problem [4.5]:

$$\begin{cases} -\dfrac{\partial u^\star}{\partial x} = -p^\star, & u^\star(a) = u_1, \\ \dfrac{\partial p^\star}{\partial x} = -\Delta_y u^\star - f, & p^\star(0) = -u_0. \end{cases}$$

4.2. Summary of results on the decoupling of optimal control problems

We will recall here a few results on the decoupling of optimal control problems for systems determined by parabolic-type partial differential equations such as those obtained by J.-L. Lions and presented in [LIO 68a]. These are the results that inspired the factorization method presented in this book.

We will work in the Hilbert setting for presenting parabolic partial differential equations. Let V and H be two Hilbert spaces such that:

$$V \subset H$$

and V is dense in H. We identify H with its dual, which gives us the classical relation:

$$V \subset H \subset V'.$$

Let A be an elliptic operator from V into V'. We will study the optimal control problem associated with the parabolic problem of evolution in time over the interval $[0, T]$ with state $u(t)$ defined by:

$$\begin{aligned} \frac{\mathrm{d}u}{\mathrm{d}t} + Au = f + Bv, \\ u(0) = u_0. \end{aligned} \qquad [4.12]$$

The right-hand side involves a constant term $f \in L^2(0, T; V')$ and the action of the control $v \in L^2(0, T; \mathcal{U})$ through the operator $B \in \mathcal{L}(\mathcal{U}, V')$, where $\mathcal{U}$, the Hilbert space identified with its dual, is the control space. This defines the state in $L^2(0, T; V)$ with $\dfrac{\mathrm{d}u}{\mathrm{d}t} \in L^2(0, T; V')$ and so $u \in \mathcal{C}(0, T; H)$.

Define $\mathcal{H}$ as the Hilbert space of observations and $C \in \mathcal{L}(V, \mathcal{H})$ as the observation operator. We write $A^\star$ the adjoint of the operator A. Define an observation $z \in L^2(0, T; \mathcal{H})$ of the state by $z(t) = Cu(t; v)$. Let z_d be the observation of the desired state, then the cost function of the control problem to be minimized is:

$$J(v) = \int_0^T \|z(t) - z_d(t)\|_{\mathcal{H}}^2 \,\mathrm{d}t + \int_0^T \|v(t)\|_{\mathcal{U}}^2 \,\mathrm{d}t. \qquad [4.13]$$

We introduce the adjoint state p as:

$$-\frac{dp}{dt} + A^{\star}p = C^{\star}(Cu - z_d),$$
$$p(T) = 0. \quad [4.14]$$

We have that:

$$p \in L^2(0,T;V), \quad \frac{dp}{dt} \in L^2(0,T;V'), \quad p \in \mathcal{C}(0,T;H).$$

The functional J is Fréchet differentiable, and, using p, its gradient can be expressed as:

$$\nabla_v J(v) = 2(v + B^{\star}p).$$

Let $v^{\star}$ be the optimal control:

$$J(v^{\star}) = \min_{v} J(v).$$

The optimal control is thus characterized by:

$$v^{\star} = -B^{\star}p^{\star},$$

where $p^{\star}$ is the adjoint state corresponding to the optimal control $v^{\star}$ and the optimal state $u^{\star}$. The optimality system can therefore be written as:

$$\frac{du^{\star}}{dt} + Au^{\star} = f - BB^{\star}p^{\star}, \quad u^{\star}(0) = u_0,$$
$$-\frac{dp^{\star}}{dt} + A^{\star}p^{\star} = C^{\star}(Cu^{\star} - z_d), \quad p^{\star}(T) = 0.$$

For simplicity, we omit the notation.* indicating optimality, and set $D_1 = BB^{\star} \in \mathcal{L}(L^2(0,T;V), L^2(0,T;V'))$, $D_2 = C^{\star}C \in \mathcal{L}(L^2(0,T;V), L^2(0,T;V'))$ and $g = -C^{\star}z_d \in L^2(0,T;V')$, so that the coupled optimality system for increasing and decreasing time is given by:

$$\frac{du}{dt} + Au + D_1p = f, \quad u(0) = u_0, \quad [4.15]$$

$$-\frac{dp}{dt} + A^{\star}p - D_2u = g, \quad p(T) = 0. \quad [4.16]$$

Bellman invariant embedding then consists of considering the family of control problems defined on the horizon $[s, T]$ with initial state $h \in H$ at $t = s$. By linearity of equations [4.15] and [4.16], the adjoint state $p(s)$ has an affine dependency on h:

$$p(s) = P(s)h + r(s). \quad [4.17]$$

J.-L. Lions showed that $P(s) \in \mathcal{L}(H, H)$, P is self-adjoint and positive definite. Furthermore, $P(s)$ can be used to describe the quadratic component of the optimal cost for the control problem on the time interval $[s, T]$ with initial state $u(s) = h$:

$$J(v^\star_{s,h}) = 2(h, P(s)h) + \text{linear terms or constants in } h. \quad [4.18]$$

The operator $P(s)$ can be used to find the optimal control given the present state $u(s)$, that is, the optimal feedback. Indeed, we have that:

$$v^\star_{s,h} = -B^\star(P(s)u(s) + r(s)). \quad [4.19]$$

The state "looped" by the optimal feedback satisfies:

$$\frac{\mathrm{d}u}{\mathrm{d}t} + (A + D_1 P)u = f - D_1 r. \quad [4.20]$$

J.-L. Lions showed that P satisfies the Riccati equation:

$$-\frac{\mathrm{d}P}{\mathrm{d}t} + PA + A^\star P + PD_1 P = D_2, \quad P(T) = 0. \quad [4.21]$$

In [LIO 68a], he established the well-posedness of this equation, which has a solution in $L^2(0, T; \mathcal{L}(H, H))$. This equation has since been studied more directly by other authors, such as [BEN 07]. In particular, P is a weak solution in the sense of definition 2.1 of [BEN 07, p. 390], by an application of theorem 3.3 [BEN 07, p. 404], since A is a variational operator:

$$-\left(\frac{\mathrm{d}P}{\mathrm{d}t}x, y\right) + (Ax, Py) + (Px, Ay) + (B^\star Px, B^\star Py) = (Cx, Cy),$$

$$\forall x, y \in V, \quad P(T) = 0. \quad [4.22]$$

The residual r satisfies:

$$-\frac{\mathrm{d}r}{\mathrm{d}t} + A^{\star}r + PD_1 r = Pf + g, \quad r(T) = 0. \qquad [4.23]$$

4.3. Summary of results of A. Bensoussan on Kalman optimal filtering

Using the same framework as Lions [LIO 68a] in his book [BEN 71], A. Bensoussan studied Kalman filtering in systems governed by parabolic equations. In a deterministic context, filtering can be reduced to a control problem. We will summarize these results here.

Consider the system with state $u(t)$ satisfying the following equations, whose initial state and right-hand side are, respectively, subject to the perturbations η and ξ:

$$\begin{aligned} &\frac{\mathrm{d}u}{\mathrm{d}t} + Au = f + \xi, \\ &\quad u(0) = \eta. \end{aligned} \qquad [4.24]$$

Filtering then consists of finding an estimate $\hat{u}$ of u at time t, given an observation $z = Cu$ and assuming that the perturbations η and ξ have minimum norms. We therefore wish to minimize the distance between the true observation z_d and the observation arising from the model $z = Cu$, while minimizing η and ξ. Thus, we define the cost function:

$$J(\xi,\eta) = \int_0^T \|z(t) - z_d(t)\|_{\mathcal{H}}^2 \,\mathrm{d}t + \int_0^T \|\xi(t)\|_H^2 \,\mathrm{d}t + \|\eta\|_H^2 . \qquad [4.25]$$

For this optimal control problem, we introduce the adjoint state p satisfying:

$$\begin{aligned} &-\frac{\mathrm{d}p}{\mathrm{d}t} + A^{\star}p = C^{\star}(Cu - z_d), \\ &\quad p(T) = 0. \end{aligned} \qquad [4.26]$$

The optimal control is characterized by:

$$\xi(t) = -p(t), \qquad [4.27]$$

$$\eta = -p(0). \qquad [4.28]$$

The coupled optimality system is, therefore:

$$\frac{du^\star}{dt} + Au^\star = f - p^\star, \qquad u^\star(0) = -p(0), \qquad [4.29]$$

$$-\frac{dp^\star}{dt} + A^\star p^\star = C^\star(Cu^\star - z_d), \qquad p^\star(T) = 0. \qquad [4.30]$$

Invariant embedding can therefore be naturally performed in the direction of increasing t. This is similar to the approach of section 1.3. Consider the restriction of [4.29], [4.30] to the interval $]0, s[$ with $p(s) = h$. We then have:

$$u(s) = Q(s)h + w(s), \qquad [4.31]$$

where $Q(s) \in \mathcal{L}(H, H)$ is self-adjoint and positive definite. Then, Q satisfies:

$$\left(\frac{dQ}{dt}x, y\right) + (A^\star x, Qy) + (Qx, A^\star y) - (CQx, CQy) = -(x, y),$$
$$\forall x, y \in V, \quad Q(0) = -I. \qquad [4.32]$$

The residual w satisfies:

$$\frac{dw}{dt} + Aw - QC^\star Cw = f - QC^\star z_d, \quad w(0) = 0. \qquad [4.33]$$

In the filtering studied in [BEN 71], this process is applied recursively by using the observations z gathered between the initial time 0 and the present time t. The initial condition of [4.30] must therefore be taken at the present time, and formula [4.31] shows that the optimal estimate of the state at the present time is given by w satisfying [4.33], which is known as the filter equation. This is not the case in our approach of backward factorization in section 1.3, but Riccati equation [4.32] is highly similar to the equation in this section.

4.4. Parabolic regularization for the factorization of elliptic boundary value problems

Now that we have summarized these results, we return to the factorization of elliptic boundary value problems in an open cylindrical set. The

x-coordinate will play the role of the time variable in the control problem. To remain consistent with both the control theory summarized in section 4.2 and the factorization presented in previous chapters, we will flip the direction of the x-axis in this analogy with respect to time: the value $x = 0$ corresponds to the final time $t = T$ and $x = a$ corresponds to the initial time $t = 0$.

If we wished to directly apply the results of section 4.2 to problem $(\mathcal{P}_0)$ as presented in section 4.1, we would take $A = 0, B = I, C = \nabla_y$, and $V = H_0^1(\mathcal{O})$. Formally, this implies $D_1 = I$ and $D_2 = -\Delta_y$, so that we recover the Riccati equation [1.22] after performing the change of variable $x' = a - x$. But, as we remarked earlier, for u satisfying [4.1], we do not have that $u \in L^2(O, T; H_0^1(\mathcal{O}))$, and the method of [LIO 68a] cannot be applied directly. We therefore propose to perform a parabolic regularization of [4.1]:

$$\begin{cases} -\dfrac{\partial u_\varepsilon}{\partial x} - \varepsilon \Delta_y u_\varepsilon = v \text{ in } \Omega, \\ u_\varepsilon(a) = u_1, \qquad\qquad u_\varepsilon|_\Sigma = 0. \end{cases} \tag{4.34}$$

This equation is indeed parabolic, and so we are in the situation of section 4.2 with $A = -\varepsilon \Delta_y$. We define the regularized control problem as the minimization of the cost function:

$$J_\varepsilon(v) = \int_0^a \|\nabla_y u_\varepsilon(v) - \nabla_y u_d\|_{L^2(\mathcal{O})}^2 \, \mathrm{d}x + \int_0^a\!\!\int_{\mathcal{O}} v^2 \, \mathrm{d}x \, \mathrm{d}y$$
$$- 2 \int_{\Gamma_0} u_0(y) u_\varepsilon(0, y; v) \, \mathrm{d}y. \tag{4.35}$$

We have that:

$$J_\varepsilon(v) = \overline{J}(u_\varepsilon).$$

The adjoint state of this problem is defined by:

$$\begin{cases} \dfrac{\partial p_\varepsilon}{\partial x} - \varepsilon \Delta_y p_\varepsilon = -\Delta_y u_\varepsilon - f \text{ in } \Omega, \\ p_\varepsilon(0) = -u_0, \qquad\qquad p_\varepsilon|_\Sigma = 0. \end{cases} \tag{4.36}$$

The optimality condition is:

$$v_\varepsilon^\star = -p_\varepsilon^\star. \tag{4.37}$$

Omitting the $\star$ notation, the optimality system is therefore given by:

$$\begin{cases} -\dfrac{\partial u_\varepsilon}{\partial x} - \varepsilon\Delta_y u_\varepsilon = -p_\varepsilon, & u_\varepsilon(a) = u_1, \quad u_\varepsilon|_\Sigma = 0, \\ \dfrac{\partial p_\varepsilon}{\partial x} - \varepsilon\Delta_y p_\varepsilon = -\Delta_y u_\varepsilon - f, & p_\varepsilon(0) = -u_0, \quad p_\varepsilon|_\Sigma = 0. \end{cases} \quad [4.38]$$

The state and adjoint state have the following regularities:

$$u_\varepsilon, p_\varepsilon \in L^2(0,a;H_0^1(\mathcal{O})); \quad \frac{\mathrm{d}u_\varepsilon}{\mathrm{d}x}, \frac{\mathrm{d}p_\varepsilon}{\mathrm{d}x} \in L^2(0,a;H^{-1}(\mathcal{O})). \quad [4.39]$$

From the results of J.-L. Lions summarized in section 4.2, this problem can be decoupled by introducing the operator $P_\varepsilon \in \mathcal{L}(L^2(\mathcal{O}); L^2(\mathcal{O}))$, such that:

$$p_\varepsilon(x) = P_\varepsilon(x)u_\varepsilon(x) + r_\varepsilon(x). \quad [4.40]$$

We obtain that:

$$\frac{\mathrm{d}P_\varepsilon}{\mathrm{d}x} - \varepsilon P_\varepsilon \Delta_y - \varepsilon\Delta_y P_\varepsilon + P_\varepsilon^2 = -\Delta_y, \quad P_\varepsilon(0) = 0, \quad [4.41]$$

which may be written in variational form using [4.22]:

$$(\frac{\mathrm{d}P_\varepsilon}{\mathrm{d}x}h, \overline{h}) - \varepsilon(\Delta_y h, P_\varepsilon\overline{h}) - \varepsilon(P_\varepsilon h, \Delta_y\overline{h}) + (P_\varepsilon h, P_\varepsilon\overline{h}) = (\nabla_y h, \nabla_y\overline{h}),$$
$$\forall h, \overline{h} \in H_0^1(\mathcal{O}), P_\varepsilon(0) = 0. \quad [4.42]$$

The decoupled equations are given by:

$$\frac{\partial r_\varepsilon}{\partial x} - \varepsilon\Delta_y r_\varepsilon + P_\varepsilon r_\varepsilon = -f, \quad r_\varepsilon(0) = -u_0, \quad r_\varepsilon|_\Sigma = 0, \quad [4.43]$$

$$-\frac{\partial u_\varepsilon}{\partial x} - \varepsilon\Delta_y u_\varepsilon + P_\varepsilon u_\varepsilon = -r_\varepsilon, \quad u_\varepsilon(a) = u_1, \quad u_\varepsilon|_\Sigma = 0. \quad [4.44]$$

We still need to study the behavior of this system as $\varepsilon \to 0$. Consider again system [4.38], and eliminate p_ε. This gives the anisotropic elliptical boundary value problem:

$$-\frac{\partial^2 u_\varepsilon}{\partial x^2} - \Delta_y u_\varepsilon + \varepsilon^2 \Delta_y^2 u_\varepsilon = f,$$

$$\left(-\frac{\partial u_\varepsilon}{\partial x} - \varepsilon \Delta_y u_\varepsilon\right)|_{\Gamma_0} = u_0, \quad u_\varepsilon(a) = u_1, \tag{4.45}$$

$$u_\varepsilon|_\Sigma = 0, \quad \Delta_y u_\varepsilon|_\Sigma = 0.$$

For the second boundary condition on Σ, we used the fact that since u_ε vanishes on Σ, its derivative w.r.t. x is also zero. We will now show that this problem is well-posed in the space $X_\varepsilon = \{u \in L^2(0, a; H_0^1(\mathcal{O})) : \dfrac{\partial u}{\partial x} + \varepsilon \Delta_y u \in L^2(\Omega)\}$. We equip X_ε with the norm:

$$\|.\|_{X_\varepsilon} = \left(\|.\|^2_{L^2(0,a;H_0^1(\mathcal{O}))} + \left\| (\frac{\partial}{\partial x} + \varepsilon \Delta_y). \right\|^2_{L^2(\Omega)} \right)^{1/2}.$$

Note that this space is dependent on ε. For $u \in X_\varepsilon$, we have that $\dfrac{\partial u}{\partial x} \in L^2(0, a; H^{-1}(\mathcal{O}))$ and therefore $u \in \mathcal{C}([0, a]; L^2(\mathcal{O}))$, and we can thus define the trace of u, $u|_{\Gamma_s}$, in $L^2(\mathcal{O})$ for $0 \leq s \leq a$. For $\varepsilon > 0$, this mapping is surjective: indeed, for $u_1 \in L^2(\mathcal{O})$ and $s = a$, there exists a lifting w in X_ε, such that $w|_{\Gamma_a} = u_1$; for example, we can choose w satisfying:

$$\frac{\partial w}{\partial x} + \varepsilon \Delta_y w = 0, \quad w|_\Sigma = 0, \quad w|_{\Gamma_a} = u_1. \tag{4.46}$$

This mapping is continuous but not uniformly continuous in ε.

PROPOSITION 4.3.– Problem [4.45] admits a unique solution in X_ε for $f \in X'_\varepsilon$, $u_0 \in L^2(\mathcal{O})$, $u_1 \in L^2(\mathcal{O})$ and for $\varepsilon > 0$.

PROOF.– Note that:

$$-\frac{\partial^2 u}{\partial x^2} + \varepsilon^2 \Delta_y^2 u = (-\frac{\partial}{\partial x} + \varepsilon \Delta_y)(\frac{\partial u}{\partial x} + \varepsilon \Delta_y u).$$

Let u and v be two regular functions satisfying the boundary conditions of X_ε on Σ. Using Green's formula [3.2] and by transposition of the operator

$(-\frac{\partial}{\partial x} + \varepsilon\Delta_y)$, we obtain:

$$\int_\Omega (-\frac{\partial^2 u}{\partial x^2} + \varepsilon^2 \Delta_y^2 u - \Delta_y u) v \, dx \, dy = \int_\Omega (\frac{\partial u}{\partial x} + \varepsilon\Delta_y u)(\frac{\partial v}{\partial x} + \varepsilon\Delta_y v) \, dx \, dy$$
$$+ \int_\Omega \nabla_y u . \nabla_y v \, dx \, dy - \int_\Sigma \varepsilon^2 \Delta_y u \frac{\partial v}{\partial n} \, d\Sigma - \int_{\Gamma_a} (\frac{\partial u}{\partial x} + \varepsilon\Delta_y u)|_{\Gamma_a} v|_{\Gamma_a} \, dy$$
$$+ \int_{\Gamma_0} (\frac{\partial u}{\partial x} + \varepsilon\Delta_y u)|_{\Gamma_0} v|_{\Gamma_0} \, dy. \qquad [4.47]$$

By denseness, we extend this formula to u_ε, a solution of [4..45] and v a test function in X_ε by adjusting the left-hand side of [4.47] by $< (-\frac{\partial^2 u_\varepsilon}{\partial x^2} + \varepsilon^2 \Delta_y^2 u_\varepsilon - \Delta_y u_\varepsilon), v >_{X_\varepsilon' \times X_\varepsilon}$. Indeed, for $v \in X_\varepsilon$, we can define the traces $v|_{\Gamma_0}$ and $v|_{\Gamma_a}$ in $L^2(\mathcal{O})$, which together with the condition on Γ_0 from [4.45] allows us to interpret the last term of [4.47]. The relation obtained with $f = 0$ and $u_0 = 0$ allows us to define $(\frac{\partial u_\varepsilon}{\partial x} + \varepsilon\Delta_y u_\varepsilon)|_{\Gamma_a}$ in $L^2(\mathcal{O})$ (or in a similar fashion on any section Γ_s).

We now consider u_ε satisfying [4.45] with the homogeneous data $u_1 = 0$ on Γ_a. Let $X_{0,\varepsilon} = \{v \in X_\varepsilon : v|_{\Gamma_a} = 0\}$. Choosing $v \in X_{0,\varepsilon}$, problem [4.45] has the variational formulation:

$$\int_\Omega (\frac{\partial u_\varepsilon}{\partial x} + \varepsilon\Delta_y u_\varepsilon)(\frac{\partial v}{\partial x} + \varepsilon\Delta_y v) \, dx \, dy + \int_\Omega \nabla_y u_\varepsilon . \nabla_y v \, dx \, dy$$
$$= \int_{\Gamma_0} u_0 v|_{\Gamma_0} \, dy + < f, v >_{X_\varepsilon' \times X_\varepsilon} . \qquad [4.48]$$

The left-hand side defines a bilinear form, continuous and coercive on $X_{0,\varepsilon}$, and the right-hand side defines a linear form continuous on $X_{0,\varepsilon}$. From theorem 3.1 in [LIO 68b], which we summarized in proposition 1.1, for $v \in X_{0,\varepsilon}$, we have:

$$\|v|_{\Gamma_0}\|_{L^2(\Gamma_0)} \leq C \left(\|v\|_{L^2(0,a;H_0^1(\mathcal{O}))} + \left\| \frac{\partial v}{\partial x} \right\|_{L^2(0,a;H^{-1}(\mathcal{O}))} \right), \qquad [4.49]$$

and, therefore:

$$\|v|_{\Gamma_0}\|_{L^2(\Gamma_0)} \leq C \left(\|v\|_{L^2(0,a;H_0^1(\mathcal{O}))} + \left\| \frac{\partial v}{\partial x} \right\|_{L^2(0,a;H^{-1}(\mathcal{O}))} \right), \quad [4.50]$$

and we note that the constant C' is independent of ε.

Existence and uniqueness of the solution of problem [4.45] in $X_{0,\varepsilon}$ now follow from the Lax–Milgram theorem.

Now consider non-zero boundary data on Γ_a, written $u_1 \in L^2(\mathcal{O})$. Let w be a lifting in X_ε of u_1 defined by [4.46].

Then, using [4.46], $\bar{u}_\varepsilon = u_\varepsilon - w$ satisfies:

$$-\frac{\partial^2 \bar{u}_\varepsilon}{\partial x^2} - \Delta_y \bar{u}_\varepsilon + \varepsilon^2 \Delta_y^2 \bar{u}_\varepsilon = f + \Delta_y w, \quad [4.51]$$

$$(-\frac{\partial \bar{u}_\varepsilon}{\partial x} - \varepsilon \Delta_y \bar{u}_\varepsilon)|_{\Gamma_0} = u_0, \quad \bar{u}_\varepsilon(a) = 0, \quad [4.52]$$

$$\bar{u}_\varepsilon|_\Sigma = 0, \quad \Delta_y \bar{u}_\varepsilon|_\Sigma = 0. \quad [4.53]$$

It then suffices to apply the previous result to $\bar{u}_\varepsilon$ to complete the proof of the proposition. From these calculations, we deduce the following estimate of u_ε:

$$\left\| \frac{\partial u_\varepsilon}{\partial x} + \varepsilon \Delta_y u_\varepsilon \right\|_{L^2(\Omega)}^2 + \|\nabla_y u_\varepsilon\|_{L^2(\Omega)}^2 \leq$$

$$\leq C \left(\|f\|_{L^2(\Omega)}^2 + \|u_0\|_{L^2(\mathcal{O})}^2 + \|\nabla_y w\|_{L^2(\Omega)}^2 \right)$$

$$\leq C \left(\|f\|_{X'_\varepsilon}^2 + \|u_0\|_{L^2(\mathcal{O})}^2 \right) + C_\varepsilon \|u_1\|_{L^2(\mathcal{O})}^2 , \quad [4.54]$$

where C is independent of ε but C_ε is not. □

REMARK 4.1.– In the definition of problem $(\mathcal{P}_0)$, we chose $f \in \hat{X}' = L^2(0,a;H^{-1}(\mathcal{O}))$ in order to implement invariant embedding, which was justified in remark 1.1. In a similar fashion here, for the invariant embedding applied in problem [4.45], it is convenient to choose $f \in L^2(0,a;H^{-2}(\mathcal{O}))$. However, since this procedure is for regularization, the right-hand side considered here is that of problem $(\mathcal{P}_0)$. It suffices to simply note that $L^2(0,a;H^{-1}(\mathcal{O})) \subset L^2(0,a;H^{-2}(\mathcal{O}))$.

In light of remark 4.1, we will now always take the right-hand side $f \in \hat{X}'$. In the following section, we will study the convergence of the operator P_ε, initially assuming that the data u_1 is in $H^{1/2}_{0,0}(\mathcal{O})$.

4.4.1. *Convergence of the operator* P_ε

As we saw earlier, the "natural" data of the Dirichlet condition for problem $(\mathcal{P}_\varepsilon)$ on Γ_a (that is, that leads to a variational formulation) are in $L^2(\mathcal{O})$, whereas for $(\mathcal{P}_0)$, these data are in $H^{1/2}_{0,0}(\mathcal{O})$. We will therefore study the additional regularity of the solution of $(\mathcal{P}_\varepsilon)$ when $u_1 \in H^{1/2}_{0,0}(\mathcal{O})$ by using suitable liftings for these data.

PROPOSITION 4.4.– We adopt the conditions of proposition 4.3. Assume further that $u_1 \in H^{1/2}_{0,0}(\mathcal{O})$. We then have the estimate:

$$\left\| \frac{\partial u_\varepsilon}{\partial x} + \varepsilon \Delta_y u_\varepsilon \right\|^2_{L^2(\Omega)} + \|\nabla_y u_\varepsilon\|^2_{L^2(\Omega)} \leq C \Big(\|f\|^2_{L^2(\Omega)} + \|u_0\|^2_{L^2(\mathcal{O})} + \|u_1\|^2_{H^{1/2}_{0,0}(\mathcal{O})} \Big), \quad [4.55]$$

where C is independent of ε.

PROOF.– For simplicity, we will use the notation introduced at the beginning of chapter 2 with $H^0 = L^2(\mathcal{O}), H^1 = H^1_0(\mathcal{O})$, so that $H^{1/2} = H^{1/2}_{0,0}(\mathcal{O})$ and $H^{3/2} = H^{3/2}(\mathcal{O}) \cap H^1_0(\mathcal{O})$.

If $u_1 \in H^{1/2}$, we choose a lifting w_ε as follows: begin by defining $\tilde{w}_\varepsilon$ as the solution of the backward parabolic problem:

$$-\frac{\partial \tilde{w}_\varepsilon}{\partial x} + A\tilde{w}_\varepsilon - \varepsilon \Delta_y \tilde{w}_\varepsilon = 0, \quad \tilde{w}_\varepsilon(a) = u_1, \quad \tilde{w}_\varepsilon|_\Sigma = 0, \qquad [4.56]$$

where $A = (-\Delta_y)^{1/2}$. This problem is well-posed as the operator $-\varepsilon\Delta_y + A$ meets the continuity and coercivity requirements on H^1.

Let ψ be an increasing C^∞ function in x such that $0 \leq \psi(x) \leq 1$, and such that ψ is 0 in a neighborhood of 0 and 1 in a neighborhood of a. We then set:

$$w_\varepsilon = \psi \tilde{w}_\varepsilon. \qquad [4.57]$$

This defines a function from X_ε with trace equal to u_1 at $x = a$ and 0 at $x = 0$. We do the same for the case $\varepsilon = 0$. Let $\tilde{w}$ be a solution of:

$$-\frac{\partial \tilde{w}}{\partial x} + A\tilde{w} = 0, \quad \tilde{w}(a) = u_1, \quad \tilde{w}|_\Sigma = 0, \qquad [4.58]$$

in a sense to be specified at a later point. We similarly set:

$$w = \psi\tilde{w}. \qquad [4.59]$$

We now apply lemma 4.1, the proof of which is deferred until the end of this demonstration.

LEMMA 4.1.– The lifting w_ε of $u_1 \in H^{1/2}$ defined by [4.56] and [4.57] converges strongly in X_ε to w defined by [4.58] and [4.59].

Writing u_ε in the form $u_\varepsilon = (u_\varepsilon - w_\varepsilon) + w_\varepsilon$ and using the variational formulation [4.48], we have the estimate:

$$\left\|\frac{\partial u_\varepsilon}{\partial x} + \varepsilon\Delta_y u_\varepsilon\right\|^2_{L^2(\Omega)} + \|\nabla_y u_\varepsilon\|^2_{L^2(\Omega)} \leq 2(\|u_\varepsilon - w_\varepsilon\|^2_{X_\varepsilon} + \|w_\varepsilon\|^2_{X_\varepsilon})$$
$$\leq C\Bigg(\|f\|^2_{L^2(\Omega)} + \|u_0\|^2_{L^2(\mathcal{O})} \qquad [4.60]$$
$$+ \left\|\frac{\partial w_\varepsilon}{\partial x} + \varepsilon\Delta_y w_\varepsilon\right\|^2_{L^2(\Omega)} + \|\nabla_y w_\varepsilon\|^2_{L^2(\Omega)}\Bigg).$$

From the convergence results in lemma 4.1, we find that:

$$\left\|\frac{\partial u_\varepsilon}{\partial x} + \varepsilon\Delta_y u_\varepsilon\right\|^2_{L^2(\Omega)} + \|\nabla_y u_\varepsilon\|^2_{L^2(\Omega)} \leq$$
$$\leq C\Bigg(\|f\|^2_{L^2(\Omega)} + \|u_0\|^2_{L^2(\mathcal{O})}$$
$$+ \left\|\frac{\partial w}{\partial x}\right\|^2_{L^2(\Omega)} + \|\nabla_y w\|^2_{L^2(\Omega)}\Bigg)$$
$$\leq C\left(\|f\|^2_{L^2(\Omega)} + \|u_0\|^2_{L^2(\mathcal{O})} + \|u_1\|^2_{H^{1/2}_{0,0}(\mathcal{O})}\right). \qquad [4.61]$$

□

We now return to the proof of lemma 4.1.

PROOF.– The operator A is continuous from H^1 to H^0 and is coercive in $H^{1/2}$. Taking $H^{1/2}$ as the pivot space, the theory of abstract parabolic equations shows that $\tilde{w}$ is defined in $L^2(0,a;H^1) \cap H^1(0,a;H^0) \subset C([0,a];H^{1/2})$. We will use the powers of the operator A. In particular, we have that:

$$A^{1/2} \in \mathcal{L}(H^{1/2}, H^0) \cap \mathcal{L}(H^1, H^{1/2}), \quad A^{3/2} \in \mathcal{L}(H^{3/2}, H^0).$$

Then, the weak formulation of [4.58] gives:

$$\int_0^a (A\tilde{w}, \frac{\partial \varphi}{\partial x})\,\mathrm{d}s + \int_0^a (A\tilde{w}, A\varphi)\,\mathrm{d}s = (A^{1/2}u_1, A^{1/2}\varphi(a)),$$

$$\forall \varphi \text{ such that } \varphi(0) = 0, \varphi \in L^2(0,a;H^1), \frac{\partial \varphi}{\partial x} \in L^2(0,a;H^0). \quad [4.62]$$

Multiplying [4.56] by $A\tilde{w}_\varepsilon$, integrating over $\mathcal{O}$ and from x to a, we obtain:

$$\frac{1}{2}\left\|A^{1/2}\tilde{w}_\varepsilon(x)\right\|^2 + \int_x^a \|A\tilde{w}_\varepsilon(s)\|^2\,\mathrm{d}s + \varepsilon \int_x^a \left\|A^{3/2}\tilde{w}_\varepsilon(s)\right\|^2 \mathrm{d}s =$$

$$= \frac{1}{2}\left\|A^{1/2}u_1\right\|^2, \quad [4.63]$$

which yields the following estimates (C represents various constants independent of ε and of u_1 in some ball of $H^{1/2}$):

$$\|\tilde{w}_\varepsilon\|_{L^\infty(0,a;H^{1/2})} \leq C, \quad \|\tilde{w}_\varepsilon\|_{L^2(0,a;H^1)} \leq C, \quad \sqrt{\varepsilon}\,\|\tilde{w}_\varepsilon\|_{L^2(0,a;H^{3/2})} \leq C.$$

By standard compactness arguments, we deduce that:

$$\tilde{w}_\varepsilon \to \tilde{\tilde{w}} \text{ in } L^\infty(0,a;H^{1/2}) \text{ weak* and in } L^2(0,a;H^1) \text{ weak,}$$

$$\varepsilon \Delta_y \tilde{w}_\varepsilon \to 0 \text{ weakly in } L^2(0,a;(H^{-1/2}))$$

and $\tilde{\tilde{w}}$ satisfies [4.62]. Hence, $\tilde{\tilde{w}} = \tilde{w}$. From equation [4.56], we therefore deduce that, as $\varepsilon \to 0$:

$$\frac{\partial \tilde{w}_\varepsilon}{\partial x} \to \frac{\partial \tilde{w}}{\partial x} \text{ in } L^2(0,a;(H^{-1/2})) \text{ weakly.} \quad [4.64]$$

After multiplying by ψ, the same result holds for w_ε and w. To establish strong convergence, we use the arguments presented in [LIO 73] (theorem 3.1 p. 106), considering the difference $\tilde{w}_\varepsilon - \tilde{w}$. To simplify the notation, we set $Z_x = \{u | u \in L^2(x, a; H_0^1(\mathcal{O})); \dfrac{\mathrm{d}u}{\mathrm{d}x} \in L^2(x, a; L^2(\mathcal{O}))\}$ and $W_x = L^2(x, a; (H_0^1(\mathcal{O}) \cap H^{3/2}(\mathcal{O}))$. By abuse of the notation (this is not a norm), we write $\|w\|_{Z_x}^2 = \frac{1}{2}\|w(x)\|_{H_{0,0}^{1/2}(\mathcal{O})}^2 + \|w\|_{L^2(x,a;H_0^1(\mathcal{O}))}^2$. The factor of $\frac{1}{2}$ will serve a purpose in future steps. Similarly, we set $(u, v)_{Z_x} = \frac{1}{2}(u(x), v(x))_{H_{0,0}^{1/2}(\mathcal{O})} + \int_x^a (Au(s), Av(s))\,\mathrm{d}s$. We have that:

$$\begin{aligned}\|\tilde{w} - \tilde{w}_\varepsilon\|_{Z_x}^2 &= \|\tilde{w}\|_{Z_x}^2 - 2(\tilde{w}, \tilde{w}_\varepsilon)_{Z_x} + \|\tilde{w}_\varepsilon\|_{Z_x}^2 \\ &\leq \varepsilon\,\|\tilde{w}_\varepsilon\|_{W_x}^2 + \|\tilde{w}\|_{Z_x}^2 - 2(\tilde{w}, \tilde{w}_\varepsilon)_{Z_x} + \|\tilde{w}_\varepsilon\|_{Z_x}^2. \qquad [4.65]\end{aligned}$$

By a calculation similar to the previous estimate on $\tilde{w}_\varepsilon$, we choose a test function v from $W_x \cap Z_x$, then multiply [4.56] by Av, and integrate over $\mathcal{O}$ and from x to a. We then proceed to integrate by parts w.r.t. x, but only on half of the term with a derivative in x. This will allow us to take limits at the end of calculations using only the weak convergence that has already been established. We obtain:

$$\begin{aligned}&\frac{1}{2}(\tilde{w}_\varepsilon(x), v(x))_{H_{0,0}^{1/2}(\mathcal{O})} - \frac{1}{2}\int_x^a \left[(\frac{\mathrm{d}\tilde{w}_\varepsilon}{\mathrm{d}x}(s), Av(s)) - (\tilde{w}_\varepsilon(s), A\frac{\mathrm{d}v}{\mathrm{d}x}(s)) \right] \mathrm{d}s \\ &\quad + \int_x^a (A\tilde{w}_\varepsilon(s), Av(s))\,\mathrm{d}s + \varepsilon \int_x^a (A^{3/2}\tilde{w}_\varepsilon(s), A^{3/2}v(s))\,\mathrm{d}s = \\ &\quad = \frac{1}{2}(u_1, v(a))_{H_{0,0}^{1/2}(\mathcal{O})}, \qquad [4.66]\end{aligned}$$

which we rewrite as:

$$\begin{aligned}\varepsilon(\tilde{w}_\varepsilon, v)_{W_x} + (\tilde{w}_\varepsilon, v)_{Z_x} &= \frac{1}{2}(u_1, v(a))_{H_{0,0}^{1/2}(\mathcal{O})} \\ &- \frac{1}{2}\int_x^a \left[(\tilde{w}_\varepsilon(s), A\frac{\mathrm{d}v}{\mathrm{d}x}(s)) - (\frac{\mathrm{d}\tilde{w}_\varepsilon}{\mathrm{d}x}(s), Av(s)) \right] \mathrm{d}s. \qquad [4.67]\end{aligned}$$

The same calculation for $\tilde{w}$ gives:

$$(\tilde{w}, v)_{Z_x} = \frac{1}{2}(u_1, v(a))_{H^{1/2}_{0,0}(\mathcal{O})} - \frac{1}{2}\int_x^a \left[(\tilde{w}(s), A\frac{\mathrm{d}v}{\mathrm{d}x}(s)) - (\frac{\mathrm{d}\tilde{w}}{\mathrm{d}x}(s), Av(s))\right] \mathrm{d}s. \quad [4.68]$$

We can rewrite [4.67] in the following form:

$$\varepsilon \left\|\tilde{w}_\varepsilon\right\|^2_{W_x} + \left\|\tilde{w}_\varepsilon\right\|^2_{Z_x} = \varepsilon(\tilde{w}_\varepsilon, \tilde{w}_\varepsilon - v)_{W_x} + (\tilde{w}_\varepsilon, \tilde{w}_\varepsilon - v)_{Z_x} + \frac{1}{2}(u_1, v(a))_{H^{1/2}_{0,0}(\mathcal{O})} - \frac{1}{2}\int_x^a \left[(\tilde{w}_\varepsilon(s), A\frac{\mathrm{d}v}{\mathrm{d}x}(s)) - (\frac{\mathrm{d}\tilde{w}_\varepsilon}{\mathrm{d}x}(s), Av(s))\right] \mathrm{d}s.$$

Reusing $(u_1, v(a))_{H^{1/2}_{0,0}(\mathcal{O})}$ from [4.68]:

$$\varepsilon \left\|\tilde{w}_\varepsilon\right\|^2_{W_x} + \left\|\tilde{w}_\varepsilon\right\|^2_{Z_x} = \varepsilon(\tilde{w}_\varepsilon, \tilde{w}_\varepsilon - v)_{W_x} + (\tilde{w}_\varepsilon, \tilde{w}_\varepsilon - v)_{Z_x} + (\tilde{w}, v)_{Z_x} -\frac{1}{2}\int_x^a \left[((\tilde{w}_\varepsilon(s) - \tilde{w}(s)), A\frac{\mathrm{d}v(s)}{\mathrm{d}x}) - (\frac{\mathrm{d}(\tilde{w}_\varepsilon(s) - \tilde{w}(s))}{\mathrm{d}x}, Av(s))\right] \mathrm{d}s.$$

Substituting this new expression into [4.65] and choosing $v' = \tilde{w}_\varepsilon - v$:

$$\|\tilde{w} - \tilde{w}_\varepsilon\|^2_{Z_x} \leq \|\tilde{w}\|^2_{Z_x} - 2(\tilde{w}, \tilde{w}_\varepsilon)_{Z_x} + \varepsilon(\tilde{w}_\varepsilon, v')_{W_x} + (\tilde{w}_\varepsilon, v')_{Z_x} + (\tilde{w}, \tilde{w}_\varepsilon - v')_{Z_x} - \frac{1}{2}\int_x^a \left[((\tilde{w}_\varepsilon - \tilde{w}), A\frac{\mathrm{d}(\tilde{w}_\varepsilon - v')}{\mathrm{d}x}) - (\frac{\mathrm{d}(\tilde{w}_\varepsilon - \tilde{w})}{\mathrm{d}x}, A(\tilde{w}_\varepsilon - v'))\right] \mathrm{d}s. \quad [4.69]$$

Denote R_ε to the last term of the right-hand side of [4.69]. Developing gives:

$$R_\varepsilon = -\frac{1}{2}\int_x^a \left[(\tilde{w}, A\frac{\mathrm{d}v'}{\mathrm{d}x}) - (\frac{\mathrm{d}\tilde{w}}{\mathrm{d}x}, Av') + (\frac{\mathrm{d}\tilde{w}_\varepsilon}{\mathrm{d}x}, Av') - (\tilde{w}_\varepsilon, A\frac{\mathrm{d}v'}{\mathrm{d}x}) + (\frac{\mathrm{d}\tilde{w}}{\mathrm{d}x}, A\tilde{w}_\varepsilon) - (\tilde{w}, A\frac{\mathrm{d}\tilde{w}_\varepsilon}{\mathrm{d}x})\right] \mathrm{d}s. \quad [4.70]$$

But v' is an arbitrary function of $W_x \cap Z_x$ that we may fix before taking limits:

$$R_\varepsilon \to 0, \text{ as } \varepsilon \to 0. \qquad [4.71]$$

Similarly, we can take the limit of the right-hand side of [4.69], so that:

$$\|\tilde{w} - \tilde{w}_\varepsilon\|_{Z_x} \to 0,$$

which holds for all x. Multiply by ψ to cancel the $\tilde{.}$. We thus obtain:

$$\begin{aligned} w_\varepsilon \to w \quad & L^2(0,a;H_0^1(\mathcal{O})) \text{ strongly}, \\ \frac{\partial w_\varepsilon}{\partial x} + \varepsilon \Delta_y w_\varepsilon \to \frac{\partial w}{\partial x} \quad & L^2(\Omega) \text{ strongly as } \varepsilon \to 0. \end{aligned}$$

□

It is possible to show additional regularity properties in the solution of problem [4.45] if the data u_0, u_1, f are more regular. Let $X_\varepsilon^1 = \{u \in L^2(0,a;H_0^1(\mathcal{O})) : \frac{\partial u}{\partial x} + \varepsilon \Delta_y u \in L^2(0,a;H_0^1(\mathcal{O})), \Delta_y u \in L^2(\Omega)\}$.

PROPOSITION 4.5.– Problem [4.45] has a unique solution in X_ε^1 for $f \in L^2(\Omega), u_0 \in H_0^1(\mathcal{O}), u_1 \in H_0^1(\mathcal{O})$ and $\varepsilon > 0$.

PROOF.– As before, we first consider the case $u_1 = 0$ and let $X_{0,\varepsilon}^1 = \{v \in X_\varepsilon^1 : v|_{\Gamma_a} = 0\}$. In the variational formulation [4.48], we take the test function $-\Delta_y v \in X_{0,\varepsilon}$, and so for $v \in X_{0,\varepsilon}^1$ we have, in particular, that:

$$\begin{aligned} & \int_\Omega \nabla_y \left(\frac{\partial u_\varepsilon}{\partial x} + \varepsilon \Delta_y u_\varepsilon\right) . \nabla_y \left(\frac{\partial v}{\partial x} + \varepsilon \Delta_y v\right) \, dx \, dy + \int_\Omega \Delta_y u_\varepsilon \Delta_y v \, dx \, dy \\ & = \int_{\Gamma_0} \nabla_y u_0 . \nabla_y v|_{\Gamma_0} \, dy + \int_\Omega f(-\Delta_y v) \, dx \, dy. \end{aligned} \qquad [4.72]$$

This defines a bilinear form and a linear form on $X_{0,\varepsilon}^1$. Applying the Lax–Milgram theorem once again shows that problem [4.45] is well-posed in $X_{0,\varepsilon}^1$ for $u_0 \in H_0^1(\mathcal{O})$. Now, if we have a condition on Γ_a, $u_1 \in H_0^1(\mathcal{O})$, we begin by finding a lifting $w \in X_\varepsilon^1$ such that $w|_{\Gamma_a} = u_1$, then we use the same argument as above with $u_\varepsilon - w$.

Summarizing, we have the estimate:

$$\left\| \frac{\partial u_\varepsilon}{\partial x} + \varepsilon \Delta_y u_\varepsilon \right\|^2_{L^2(0,a;H_0^1(\mathcal{O}))} + \|\Delta_y u_\varepsilon\|^2_{L^2(\Omega)}$$

$$\leq C \left(\|f\|^2_{L^2(\Omega)} + \|u_0\|^2_{H_0^1(\mathcal{O})} + \|u_1\|^2_{H_0^1(\mathcal{O})} \right). \qquad [4.73]$$

□

Consider now the convergence of u_ε as $\varepsilon \to 0$. First, recall the variational formulation of problem $(\mathcal{P}_0)$: for all $v \in X_{u_1}$, $u \in X_{u_1}$ satisfies:

$$\left(\frac{\partial u}{\partial x}, \frac{\partial (v-u)}{\partial x} \right) + (\nabla_y u, \nabla_y (v-u)) = \int_{\Gamma_0} u_0 (v-u)|_{\Gamma_0} \, \mathrm{d}y$$

$$+ \int_\Omega f(v-u) \, \mathrm{d}x \, \mathrm{d}y \qquad [4.74]$$

THEOREM 4.1.– We assume that the hypotheses of proposition 4.3 are satisfied and that u_1 is in the space $H^{1/2}_{0,0}(\mathcal{O})$. Let u be the solution of problem $(\mathcal{P}_0)$ and u_ε the solution of [4.45]. We have the following convergence as ε tends to 0:

$$u_\varepsilon \to u \qquad \text{in } L^2(0,a;H_0^1(\mathcal{O})),$$

$$\frac{\partial u_\varepsilon}{\partial x} + \varepsilon \Delta_y u_\varepsilon \to \frac{\partial u}{\partial x} \qquad \text{in } L^2(\Omega).$$

PROOF.– From [4.60], we have the following estimate:

$$\left\| \frac{\partial u_\varepsilon}{\partial x} + \varepsilon \Delta_y u_\varepsilon \right\|^2_{L^2(\Omega)} + \|u_\varepsilon\|^2_{L^2(0,a;H_0^1(\mathcal{O}))} \leq C. \qquad [4.75]$$

We can thus extract a sequence $\{\varepsilon_n\} \subset \mathbb{R}^+$ such that as $n \to +\infty$, $\varepsilon_n \to 0$ and:

$$u_{\varepsilon_n} \to w \qquad \text{in } L^2(0,a;H_0^1(\mathcal{O})) \text{ weakly}, \qquad [4.76]$$

$$\frac{\partial u_{\varepsilon_n}}{\partial x} + \varepsilon_n \Delta_y u_{\varepsilon_n} = p_{\varepsilon_n} \to p \qquad \text{in } L^2(\Omega) \text{ weakly}. \qquad [4.77]$$

From this last convergence, we deduce that $p = \dfrac{\partial w}{\partial x}$ in the sense of distributions, because:

$$\begin{aligned} < p, \varphi > &= \lim_{n \to \infty} < p_{\varepsilon_n}, \varphi > \quad \forall \varphi \in \mathcal{D}(\Omega) \\ &= \lim_{n \to \infty} < u_{\varepsilon_n}, -\frac{\partial \varphi}{\partial x} + \varepsilon_n \Delta_y \varphi > = < w, -\frac{\partial \varphi}{\partial x} > . \end{aligned} \quad [4.78]$$

We can use the variational formulation [4.48] to show that w is the unique solution u of problem $(\mathcal{P}_0)$ and so these convergence statements hold without passing to a subsequence. Our proof of strong convergence is inspired by the proof of theorem 3.2 in [LIO 73, p. 109]. Let:

$$\mathcal{X}_\varepsilon = \left\| \frac{\partial (u_\varepsilon - u)}{\partial x} + \varepsilon \Delta_y u_\varepsilon \right\|^2_{L^2(\Omega)} + \|u_\varepsilon - u\|^2_{L^2(0,a;H^1_0(\mathcal{O}))} .$$

We transform this quantity by performing the following calculation: for any test function $v \in X_{u_1}$, we can construct $v_\varepsilon \in X_{u_1,\varepsilon} = \{v \in X_\varepsilon | v|_{\Gamma_a} = u_1\}$, such that:

$$\|\nabla_y (v_\varepsilon - v)\|_{L^2(\Omega)} + \left\| \frac{\partial v_\varepsilon}{\partial x} + \varepsilon \Delta_y v_\varepsilon - \frac{\partial v}{\partial x} \right\|_{L^2(\Omega)} \to 0 \text{ as } \varepsilon \to 0.$$

We introduce v_ε into the expression of $\mathcal{X}_\varepsilon$:

$$\begin{aligned} \mathcal{X}_\varepsilon = &\left(\frac{\partial (u_\varepsilon - v_\varepsilon)}{\partial x} + \varepsilon \Delta_y (u_\varepsilon - v_\varepsilon), \frac{\partial u_\varepsilon}{\partial x} + \varepsilon \Delta_y u_\varepsilon \right) \\ &+ \left(\frac{\partial v_\varepsilon}{\partial x} + \varepsilon \Delta_y v_\varepsilon, \frac{\partial u_\varepsilon}{\partial x} + \varepsilon \Delta_y u_\varepsilon \right) \\ &- 2 \left(\frac{\partial u_\varepsilon}{\partial x} + \varepsilon \Delta_y u_\varepsilon, \frac{\partial u}{\partial x} \right) + \left\| \frac{\partial u}{\partial x} \right\|^2 + \|u_\varepsilon - u\|^2_{L^2(0,a;H^1_0(\mathcal{O}))} . \end{aligned} \quad [4.79]$$

Using the variational formulation [4.48] with the test function $u_\varepsilon - v_\varepsilon$, we have that:

$$\begin{aligned}\mathcal{X}_\varepsilon = &\int_{\Gamma_0} u_0(u_\varepsilon - v_\varepsilon)|_{\Gamma_0} \, \mathrm{d}y + \int_\Omega f(u_\varepsilon - v_\varepsilon) \, \mathrm{d}x \, \mathrm{d}y \\ &- \int_\Omega \nabla_y u_\varepsilon . \nabla_y (u_\varepsilon - v_\varepsilon) \, \mathrm{d}x \, \mathrm{d}y \\ &+ \left(\frac{\partial v_\varepsilon}{\partial x} + \varepsilon \Delta_y v_\varepsilon, \frac{\partial u_\varepsilon}{\partial x} + \varepsilon \Delta_y u_\varepsilon\right) - 2\left(\frac{\partial u_\varepsilon}{\partial x} + \varepsilon \Delta_y u_\varepsilon, \frac{\partial u}{\partial x}\right) \\ &+ \left\|\frac{\partial u}{\partial x}\right\|^2 + \|u_\varepsilon - u\|^2_{L^2(0,a;H_0^1(\mathcal{O}))} .\end{aligned} \quad [4.80]$$

By regrouping the third and seventh terms of the above expression, we find that:

$$\begin{aligned}\mathcal{X}_\varepsilon = &\int_{\Gamma_0} u_0(u_\varepsilon - v_\varepsilon)|_{\Gamma_0} \, \mathrm{d}y \\ &+ \int_\Omega f(u_\varepsilon - v_\varepsilon) \, \mathrm{d}x \, \mathrm{d}y + \left(\frac{\partial v_\varepsilon}{\partial x} + \varepsilon \Delta_y v_\varepsilon, \frac{\partial u_\varepsilon}{\partial x} + \varepsilon \Delta_y u_\varepsilon\right) \\ &- 2\left(\frac{\partial u}{\partial x}, \frac{\partial u_\varepsilon}{\partial x} + \varepsilon \Delta_y u_\varepsilon\right) \\ &- 2\left(\nabla_y u_\varepsilon, \nabla_y u\right) + \left\|\frac{\partial u}{\partial x}\right\|^2 + \|\nabla_y u\|^2 + (\nabla_y u_\varepsilon, \nabla_y v_\varepsilon).\end{aligned} \quad [4.81]$$

From the convergence statements [4.76] and [4.77] and by construction of v_ε, we have that:

$$(u_\varepsilon - v_\varepsilon)|_{\Gamma_0} \to (u - v)|_{\Gamma_0} \quad \text{in } L^2(\mathcal{O}) \quad \text{weakly.} \quad [4.82]$$

Thus, by taking the limit:

$$\begin{aligned}\lim_{\varepsilon \to 0} \mathcal{X}_\varepsilon = &\int_{\Gamma_0} u_0(u - v)|_{\Gamma_0} \, \mathrm{d}y + \int_\Omega f(u - v) \, \mathrm{d}x \, \mathrm{d}y + \left(\frac{\partial u}{\partial x}, \frac{\partial (v - u)}{\partial x}\right) \\ &+ (\nabla_y u, \nabla_y (v - u)) = 0,\end{aligned} \quad [4.83]$$

which proves the desired result. □

By applying the previous theorem to the sub-cylinder Ω_s for almost every $s \in]0,a[$, we obtain the following result on Γ_s.

COROLLARY 4.1.– For $f = 0$ and $u_0 = 0$, $u|_{\Gamma_s}$ given in $H^{1/2}_{0,0}(\mathcal{O})$, as $\varepsilon \to 0$, we have that:

$$(\frac{\partial u_\varepsilon}{\partial x} + \varepsilon \Delta_y u_\varepsilon)|_{\Gamma_s} = P_\varepsilon(s) u|_{\Gamma_s} \to \frac{\partial u}{\partial x}|_{\Gamma_s} = P(s) u|_{\Gamma_s} \qquad [4.84]$$

in $(H^{1/2}_{0,0}(\mathcal{O}))'$. Therefore:

$$P_\varepsilon(s) \to P(s), \text{ in } \mathcal{L}(H^{1/2}_{0,0}(\mathcal{O}), (H^{1/2}_{0,0}(\mathcal{O})') \quad \forall s \in [0,a], \qquad [4.85]$$

and:

$$P_\varepsilon(.) \to P(.), \text{ in } \mathcal{L}(L^2(0,a; H^1_0(\mathcal{O})), L^2(0,a; L^2(\mathcal{O}))), \qquad [4.86]$$

in the sense of strong pointwise convergence.

PROOF.– Given the hypotheses of the corollary, we can define $(\frac{\partial u_\varepsilon}{\partial x} + \varepsilon \Delta_y u_\varepsilon)|_{\Gamma_s}$ in $L^2(\mathcal{O})$ using Green's formula [4.47] as:

$$\int_{\Gamma_s} (\frac{\partial u_\varepsilon}{\partial x} + \varepsilon \Delta_y u_\varepsilon)|_{\Gamma_s} v|_{\Gamma_s} \, dy = \int_{\Omega_s} (\frac{\partial u_\varepsilon}{\partial x} + \varepsilon \Delta_y u_\varepsilon)(\frac{\partial v}{\partial x} + \varepsilon \Delta_y v) \, dx \, dy + \int_{\Omega_s} \nabla_y u_\varepsilon . \nabla_y v \, dx \, dy. \qquad [4.87]$$

Given a test function $v \in X$, the right-hand side of [4.87] defines a continuous linear form on $L^2(\mathcal{O})$, as in particular it depends only on the value of v on Γ_s.

For $u|_{\Gamma_s} \in H^{1/2}_{0,0}(\mathcal{O})$, the result then follows from theorem 4.1, as taking the limit on the right-hand side of [4.87] gives:

$$\int_{\Omega_s} \frac{\partial u}{\partial x} \frac{\partial v}{\partial x} \, dx \, dy + \int_{\Omega_s} \nabla_y u . \nabla_y v \, dx \, dy = \int_{\Gamma_s} \frac{\partial u}{\partial x}|_{\Gamma_s} v|_{\Gamma_s} \, dy. \qquad [4.88]$$

Furthermore, if, for almost every $s \in]0,a[$, we apply P_ε to the solution u_ε of [4.45], we obtain $\frac{\partial u_\varepsilon}{\partial x} + \varepsilon \Delta_y u_\varepsilon$, which by theorem 4.1 converges to $\frac{\partial u}{\partial x}$ in $L^2(0,a; L^2(\mathcal{O}))$. □

REMARK 4.2.– In the above, we assumed that $u_0 \in L^2(\mathcal{O})$, whereas the cost function J defined in [4.4] is still well defined for $u_0 \in (H^{1/2}_{0,0}(\mathcal{O}))'$. We can observe that the convergence of P_ε to P is independent of this type of regularity. It is only for the study of the residual r that it is convenient to define r_ε with a regularized initial value $u_{0,\varepsilon} \in L^2(\mathcal{O})$ converging to u_0 if $u_0 \in (H^{1/2}_{0,0}(\mathcal{O}))'$.

THEOREM 4.2.– The operator $P(x)$ satisfies the Riccati equation:

$$\begin{gathered}(\frac{\mathrm{d}P}{\mathrm{d}x}h,\overline{h}) + (Ph, P\overline{h}) = (\nabla_y h, \nabla_y \overline{h}), \quad \forall h, \overline{h} \in H^1_0(\mathcal{O}), \\ P(0) = 0.\end{gathered} \qquad [4.89]$$

PROOF.– By strong convergence (4.85), we can take the limit in [4.42], from which we obtain [4.89]. □

4.4.2. *Parabolic regularization for the Neumann-to-Dirichlet operator*

We will now show how to justify the backward factorization presented in section 1.3, which leads us to apply a Neumann-to-Dirichlet operator on the mobile boundary based on the calculations performed by Alain Bensoussan on Kalman filtering summarized in section 4.3. As in section 4.4, these filtering calculations may be viewed as a parabolic regularization of the calculations for backward factorization.

The invariant embedding from section 1.3 leads to the family of problems:

$$(\tilde{\mathcal{P}}_{s,h}) \begin{cases} -\Delta u_s = f \quad \text{in } \widetilde{\Omega}_s, \\ u_s|_\Sigma = 0, \quad -\dfrac{\partial u_s}{\partial x}|_{\Gamma_s} = h, \quad u_s|_{\Gamma_a} = u_1, \end{cases}$$

where $\widetilde{\Omega}_s =]s, a[\times \mathcal{O}$, and where h is an arbitrary data from $(H^{1/2}_{0,0}(\mathcal{O}))'$. This problem may be written in the form of a control problem optimality system as follows:

$$\begin{cases} -\dfrac{\partial u}{\partial x} = -p \quad \text{in } \widetilde{\Omega}_s, \qquad u(a) = u_1, \\ \dfrac{\partial p}{\partial x} = -\Delta_y u - f \quad \text{in } \widetilde{\Omega}_s, \; -p(s) = h, \end{cases} \qquad [4.90]$$

which may be decoupled by introducing relation [1.32]:

$$u(s) = Q(s)h + r(s). \qquad [4.91]$$

We regularize system [4.90] by the system of coupled parabolic equations:

$$\begin{cases} -\dfrac{\partial u_\varepsilon}{\partial x} - \varepsilon\Delta_y u_\varepsilon = -p_\varepsilon \text{ in } \widetilde{\Omega}s, & u_\varepsilon(a) = u_1, \\ \dfrac{\partial p_\varepsilon}{\partial x} - \varepsilon\Delta_y p_\varepsilon = -\Delta_y u_\varepsilon - f \text{ in } \widetilde{\Omega}s, & -p_\varepsilon(s) = h. \end{cases} \qquad [4.92]$$

The decoupling leading to the Kalman filter studied by A. Bensoussan, which is summarized in section 4.3, may be applied for $\varepsilon > 0$. For system [4.90], decoupling is achieved by means of the operator Q_ε, which is such that $Q_\varepsilon(s) \in \mathcal{L}(L^2(\mathcal{O}), L^2(\mathcal{O}))$ and $Q_\varepsilon(.) \in \mathcal{L}(L^2(0,a;H_0^1(\mathcal{O})), L^2(0,a;H_0^1(\mathcal{O})))$. We have that:

$$u_\varepsilon(s) = Q\varepsilon(s)h + r_\varepsilon(s) = -Q\varepsilon(s)(\frac{\partial u_\varepsilon}{\partial x} + \varepsilon\Delta_y u_\varepsilon)(s) + r_\varepsilon(s). \quad [4.93]$$

This operator Q_ε satisfies the Riccati equation:

$$\left(\frac{\mathrm{d}Q_\varepsilon}{\mathrm{d}x}u, v\right) - \varepsilon\,(\Delta_y u, Q_\varepsilon v) - \varepsilon\,(Q_\varepsilon u, \Delta_y v) - (\nabla_y Q_\varepsilon u, \nabla_y Q_\varepsilon v) = -(u,v),$$
$$\forall u, v \in H_0^1(\mathcal{O}), \quad Q_\varepsilon(a) = 0. \qquad [4.94]$$

This equation is well-posed, and we will study the convergence of Q_ε to Q. It is important to note that, unlike the other situations discussed in this book where the Dirichlet-to-Neumann and Neumann-to-Dirichlet approaches are equivalent, in this case the study of the convergence of Q_ε is much simpler than that of P_ε considered in section 4.4.1. In fact, constructing a lifting and studying its convergence is not necessary here.

To study the operator Q_ε, we consider the version of system [4.90] homogeneous in h, that is, setting $f = 0$ and $u_1 = 0$. The following result can be proven in the same way as theorem 4.1, except that constructing a lifting is not required, because the only data associated with the problem are the Neumann condition on Γ_s.

THEOREM 4.3.– Let u be the solution of problem $(\mathcal{P}_0)$ and u_ε the solution of [4.45], both corresponding to the case where $f = 0$, $u_1 = 0$ and $u_0 \in L^2(\mathcal{O})$. We have the following convergence as ε tends to 0:

$$u_\varepsilon \to u \qquad \text{in } L^2(0,a;H_0^1(\mathcal{O})),$$

$$\frac{\partial u_\varepsilon}{\partial x} + \varepsilon \Delta_y u_\varepsilon \to \frac{\partial u}{\partial x} \qquad \text{in } L^2(\Omega).$$

PROOF.– From the calculation in [4.48] with $u_0 \in L^2(\mathcal{O})$ and additionally $f = 0$:

$$\int_\Omega (\frac{\partial u_\varepsilon}{\partial x} + \varepsilon \Delta_y u_\varepsilon)(\frac{\partial v}{\partial x} + \varepsilon \Delta_y v)\ \mathrm{d}x\ \mathrm{d}y + \int_\Omega \nabla_y u_\varepsilon . \nabla_y v\ \mathrm{d}x\ \mathrm{d}y$$
$$= \int_{\Gamma_0} u_0 v|_{\Gamma_0}\ \mathrm{d}y, \qquad [4.95]$$

for $v \in X_0$. Taking $v = u_\varepsilon$, we deduce that:

$$\left\| \frac{\partial u_\varepsilon}{\partial x} + \varepsilon \Delta_y u_\varepsilon \right\|^2_{L^2(\Omega)} + \|u_\varepsilon\|^2_{L^2(0,a;H_0^1(\mathcal{O}))} \leq C \|u_0\|^2_{L^2(\mathcal{O})} . \qquad [4.96]$$

Therefore, as in the proof of theorem 4.1:

$$u_\varepsilon \to u \qquad \text{in } L^2(0,a;H_0^1(\mathcal{O})) \text{ weakly}$$
$$\frac{\partial u_\varepsilon}{\partial x} + \varepsilon \Delta_y u_\varepsilon = p_\varepsilon \to p = \frac{\partial u}{\partial x} \qquad \text{in } L^2(\Omega) \text{ weakly},$$

and so:

$$\frac{\partial u_\varepsilon}{\partial x} \to \frac{\partial u}{\partial x} \qquad \text{in } L^2(0,a;H^{-1}(\mathcal{O})) \text{ weakly}. \qquad [4.97]$$

We deduce that:

$$u_\varepsilon|_{\Gamma_0} \to u|_{\Gamma_0} \quad \text{in } L^2(\mathcal{O}) \text{ weakly}. \qquad [4.98]$$

Strong convergence is obtained by considering the following quantity:

$$\mathcal{X}_\varepsilon = \left\| \frac{\partial(u_\varepsilon - u)}{\partial x} + \varepsilon \Delta_y u_\varepsilon \right\|^2_{L^2(\Omega)} + \|u_\varepsilon - u\|^2_{L^2(0,a;H_0^1(\mathcal{O}))}$$

and the calculation in the proof of theorem 4.1 gives:

$$\mathcal{X}_\varepsilon = \int_{\Gamma_0} u_0(u_\varepsilon - u)|_{\Gamma_0}\, dy. \qquad \square$$

As stated earlier, we will apply the technique of invariant embedding to problem [4.92]. We will require a convergence result stronger than theorem 4.3 with additional regularity in the data h. We will use estimates similar to those in proposition 4.5.

THEOREM 4.4.– Let u be the solution of [4.90] and u_ε the solution of [4.92], both corresponding to the case where $f = 0$, $u_1 = 0$ and $h = -p|_{\Gamma_s} \in H_0^1(\mathcal{O})$. We have the following convergence as ε tends to 0:

$$u_\varepsilon \to u \qquad \text{in } L^2(0,a; H_0^1(\mathcal{O}) \cap H^2(\mathcal{O})),$$

$$\frac{\partial u_\varepsilon}{\partial x} + \varepsilon \Delta_y u_\varepsilon \to \frac{\partial u}{\partial x} \qquad \text{in } L^2(0,a; H_0^1(\mathcal{O})).$$

PROOF.– Expression [4.72] provides the following estimate:

$$\int_s^a \left\| \nabla_y (\frac{\partial u_\varepsilon}{\partial x} + \varepsilon \Delta_y u_\varepsilon) \right\|_{L^2(\mathcal{O})}^2 dx + \int_{\tilde{\Omega}_s} |\Delta_y u_\varepsilon|^2\, dx\, dy \le C. \quad [4.99]$$

From this, we deduce the following convergence as in theorem 4.3:

$$u_\varepsilon \to u \qquad \text{in } L^2(s,a; H_0^1(\mathcal{O}) \cap H^2(\mathcal{O})) \text{ weakly},$$

$$\frac{\partial u_\varepsilon}{\partial x} + \varepsilon \Delta_y u_\varepsilon = p_\varepsilon \to p = \frac{\partial u}{\partial x} \qquad \text{in } L^2(s,a; H_0^1(\mathcal{O})) \text{ weakly},$$

and therefore:

$$\frac{\partial u_\varepsilon}{\partial x} \to \frac{\partial u}{\partial x} \qquad \text{in } L^2(\tilde{\Omega}_s) \text{ weakly}. \quad [4.100]$$

We deduce that:

$$u_\varepsilon|_{\Gamma_s} \to u|_{\Gamma_s} \quad \text{in } H_0^1(\mathcal{O}) \text{ weakly}. \quad [4.101]$$

Strong convergence is obtained in the same way as in theorem 4.3. $\square$

COROLLARY 4.2.– The operator Q satisfies:

$$\left(\frac{\mathrm{d}Q}{\mathrm{d}x}u, v\right) - (\nabla_y Qu, \nabla_y Qv) = -(u, v), \forall u, v \in H_0^1(\mathcal{O}), \quad Q(a) = 0. \tag{4.102}$$

PROOF.– For almost every $x \in [0, a]$, relation [4.101] means that $Q_\varepsilon(x) \in \mathcal{L}(H_0^1(\mathcal{O}), H_0^1(\mathcal{O}))$ converges pointwise strongly to $Q(x)$ as $\varepsilon \to 0$ in the sense that:

$$Q_\varepsilon(x)p \to Q(x)p, \quad \forall p \in H_0^1(\mathcal{O}). \tag{4.103}$$

Also:

$$\varepsilon\Delta_y u \to 0, \text{ in } H^{-1}(\mathcal{O}). \tag{4.104}$$

We can therefore take the limit in [4.94] to obtain [4.102]. □

REMARK 4.3.– In the formula obtained in corollary 4.2, the test functions were taken in $H_0^1(\mathcal{O})$, whereas, in [1.43], the test functions are naturally in $L^2(\mathcal{O})$. Note that we can extend [4.102] by denseness to $L^2(\mathcal{O})$. This is also connected to the difference mentioned earlier between the equations on P and on Q. In both cases, using a system of parabolic equations regularizes the optimality system. But P_ε, which is a bounded operator (like Q_ε), is more regular than the operator P that it approaches (P loses a degree of regularity), whereas Q_ε is less regular than Q (which gains a degree of regularity).

5

Factorization of the Discretized Problem

5.1. Introduction and problem statement

In this chapter, we wish to show how the method of invariant embedding may be applied to discretized versions of problem $(\mathcal{P}_0)$, and demonstrate the connection between this method and the Gaussian factorization. In the Appendix, we present a general approach to understanding the Gaussian factorization from the perspective of invariant embedding. We will restrict attention to the context of matrices obtained by the discretization of elliptic boundary value problems. In the first sections, we have chosen to consider an example of discretization by finite differences, but this choice is not essential: as we will see in section 5.6, the same observations can be made for finite element methods as long as the spatial discretization is based on a family of surfaces that limit the subdomains, which in the particular example of the cylinder corresponds to the family of its sections Γ_s. We will only consider factorizations based on the Dirichlet-to-Neumann operator. The same approach is applicable to factorizations based on the Neumann-to-Dirichlet operator.

To simplify the presentation, we will work in a space of dimension $n = 2$ so that the domain Ω is a rectangle $\Omega =]0, a[\times]0, b[$. We assume that f, u_0 and u_1 are sufficiently regular and that the compatibility conditions are satisfied at the corners of the domain boundary to obtain solution $u \in \mathcal{C}^3(\Omega)$ and ensure that the solution of the discretized problem converges towards the solution of the continuous problem. Let $N > 0$ be an integer, and consider the following grid: we take the same discretization step $h = a/(N - \frac{1}{2})$ in both coordinates and assume that $ph = b$ (to simplify the notation). Then (see Figure 5.1), we define the nodes by:

$$\begin{cases} a_{i,j} = ((-1/2+i)h, jh) \text{ for } i \in \{0,...,N\} \text{ and } j \in \{0,...,p\}, \\ a_{1/2,j} = (0, jh) \text{ for } j \in \{0,...,p\}. \end{cases}$$

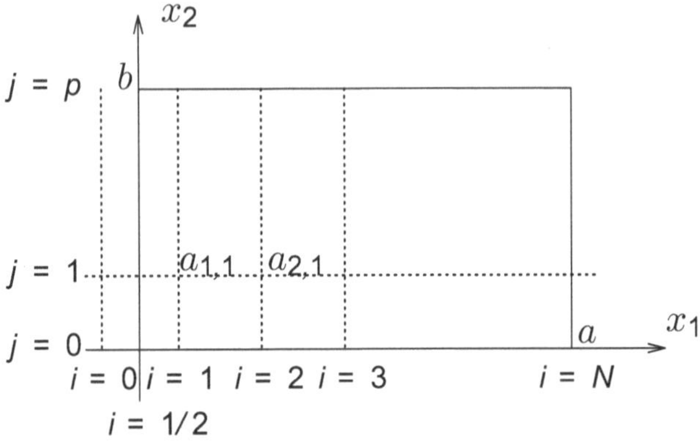

Figure 5.1. *Discretization grid of the domain*

Then, for $i \in \{2,...,N-1\}$ and $j \in \{1,...,p-1\}$, the finite differences method is based on the following approximation:

$$\begin{aligned} &-\Delta u(a_{i,j}) = f(a_{i,j}) \\ &= \frac{1}{h^2}\left(4u(a_{i,j}) - u(a_{i-1,j}) - u(a_{i+1,j}) - u(a_{i,j-1}) - u(a_{i,j+1})\right) + O(h) \end{aligned}$$

and for $i = 1$ and $j \in \{1,...,p-1\}$:

$$\begin{aligned} &-\Delta u(a_{1,j}) = f(a_{1,j}) \\ &= \frac{1}{h^2}\left(3u(a_{1,j}) - u(a_{2,j}) - u(a_{1,j-1}) - u(a_{1,j+1})\right) - \frac{1}{h}u_0(a_{1/2,j}) + O(h). \end{aligned}$$

Note that here, for $i = N$ and $j \in \{1,...,p-1\}$ (corresponding to the section Γ_a), we have $u(a_{N,j}) = u_1(a_{N,j})$, and for $i \in \{1,...,N-1\}$ and $j \in \{0,p\}$, $u(a_{i,j}) = 0$ (lateral boundary Σ). We use the notation $u(a_i) = (u(a_{i,j}))_j \in \mathbb{R}^{p-1}$ to emphasize the block tridiagonal structure of A_h :

$$u(a) = \begin{pmatrix} u(a_1) \\ \cdot \\ \cdot \\ u(a_{N-1}) \end{pmatrix} \in \mathbb{R}^{(N-1)(p-1)},$$

$$\nabla^2_{h,2} = -\frac{1}{h^2}\begin{pmatrix} 2 & -1 & & & \\ -1 & 2 & -1 & 0 & \\ & \cdot & \cdot & \cdot & \\ & & \cdot & \cdot & \cdot \\ & 0 & & \cdot & \cdot & -1 \\ & & & & -1 & 2 \end{pmatrix} \in \mathcal{M}_{(p-1)\times(p-1)}(\mathbb{R})$$

and we write I for the identity matrix in $\mathcal{M}_{(p-1)\times(p-1)}(\mathbb{R})$. Then, setting $B_1 = I - h^2\nabla^2_{h,2}$ and $B_i = 2I - h^2\nabla^2_{h,2}$ for $i = 2, ..., N-1$, we have $A_h u(a) = F_h + O(h)$, where:

$$A_h = \frac{1}{h^2}\begin{pmatrix} B_1 & -I & & & \\ -I & B_2 & -I & & 0 \\ & \cdot & \cdot & \cdot & \\ & & \cdot & \cdot & \cdot \\ & 0 & & -I & \cdot & -I \\ & & & & -I & B_{N-1} \end{pmatrix} \in \mathcal{M}_{(N-1)(p-1)\times(N-1)(p-1)}(\mathbb{R}),$$

[5.1]

$$F_h = \begin{pmatrix} F_1 \\ \cdot \\ \cdot \\ \cdot \\ F_{N-1} \end{pmatrix} = \begin{pmatrix} f_1 + \frac{u_0(a_{1/2})}{h} \\ f_2 \\ \cdot \\ f_{N-2} \\ f_{N-1} + \frac{u_1(a_N)}{h^2} \end{pmatrix} \in \mathbb{R}^{(N-1)(p-1)},$$

with $f_i = (f(a_{i,j}))_j \in \mathbb{R}^{p-1}$, for all $i \in \{1, ..., N-1\}$, $u_0(a_{1/2}) = (u_0(a_{1/2,j}))_j \in \mathbb{R}^{p-1}$ and $u_1(a_N) = (u_1(a_{N,j}))_j \in \mathbb{R}^{p-1}$. At this point, note that, if $u \in \mathcal{C}^4(\Omega)$, using this fact in Taylor's formula, we find that $A_h u(a) = F_h + O(h^2)$ (see section 3.2 in [CIA 98]). Now, the discretization by finite differences $(\mathcal{P}_h)$ of problem $(\mathcal{P}_0)$ requires us to find:

$$u_h = \begin{pmatrix} u_1 \\ \cdot \\ \cdot \\ u_{N-1} \end{pmatrix} \in \mathbb{R}^{(N-1)(p-1)},$$

such that

$$(\mathcal{P}_h) \qquad A_h u_h = F_h.$$

Proof of the convergence of u_h to u is given in [CIA 98]: for example, if $u \in \mathcal{C}^4(\Omega)$, for all $i \in \{1,...,N-1\}$ and $j \in \{1,...,p-1\}$, then $\| u_h(a_{i,j}) - u(a_{i,j}) \|_\infty = O(h^2)$.

5.2. Application of the factorization method to problem $(\mathcal{P}_h)$

Following the same procedure as in section 1.2, for each $i_0 \in \{2,...,N\}$, we define a problem similar to $(\mathcal{P}_h)$ on the subdomain $1 \le i \le i_0$ with the Dirichlet data u_{i_0} at $i = i_0$. Let $\gamma \in \mathbb{R}^{(i_0-1)(p-1)}$ be the component of u_h depending linearly on the data u_{i_0}, namely:

$$\gamma = \begin{pmatrix} \gamma_1 \\ \cdot \\ \cdot \\ \gamma_{i_0-1} \end{pmatrix} \in \mathbb{R}^{(i_0-1)(p-1)},$$

satisfying the following problem homogeneous with respect to the data u_{i_0}:

$$\frac{1}{h^2}\begin{pmatrix} B_1 & -I & & & \\ -I & B_2 & -I & 0 & \\ & \cdot & \cdot & \cdot & \\ & & \cdot & \cdot & \cdot \\ & 0 & -I & \cdot & -I \\ & & & -I & B_{i_0-1} \end{pmatrix}\begin{pmatrix} \gamma_1 \\ \cdot \\ \cdot \\ \cdot \\ \gamma_{i_0-2} \\ \gamma_{i_0-1} \end{pmatrix} = \begin{pmatrix} 0 \\ \cdot \\ \cdot \\ \cdot \\ 0 \\ \frac{u_{i_0}}{h^2} \end{pmatrix}. \qquad [5.2]$$

For each $i_0 \in \{2,...,N\}$, we also define the vector $w_{i_0-1/2} \in \mathbb{R}^{p-1}$ by $w_{i_0-1/2} = \frac{-\beta_{i_0-1}}{h}$, where β_{i_0-1} is the (i_0-1)-th component of the vector:

$$\beta = \begin{pmatrix} \beta_1 \\ \cdot \\ \cdot \\ \beta_{i_0-1} \end{pmatrix} \in \mathbb{R}^{(i_0-1)(p-1)},$$

which satisfies:

$$\frac{1}{h^2}\begin{pmatrix} B_1 & -I & & & \\ -I & B_2 & -I & 0 & \\ & \cdot & \cdot & \cdot & \\ & & \cdot & \cdot & \cdot \\ & 0 & -I & \cdot & -I \\ & & & -I & B_{i_0-1} \end{pmatrix}\begin{pmatrix} \beta_1 \\ \cdot \\ \cdot \\ \cdot \\ \beta_{i_0-2} \\ \beta_{i_0-1} \end{pmatrix} = \begin{pmatrix} f_1 + \frac{u_0(a_{1/2})}{h} \\ f_2 \\ \cdot \\ \cdot \\ f_{i_0-2} \\ f_{i_0-1} \end{pmatrix}.$$

Then, the vector $u = \gamma + \beta$ satisfies:

$$\frac{1}{h^2}\begin{pmatrix} B_1 & -I & & & & \\ -I & B_2 & -I & & 0 & \\ & \cdot & \cdot & \cdot & & \\ & & \cdot & \cdot & \cdot & \\ & 0 & & -I & \cdot & -I \\ & & & & -I & B_{i_0-1} \end{pmatrix}\begin{pmatrix} u_1 \\ \cdot \\ \cdot \\ \cdot \\ \cdot \\ u_{i_0-1} \end{pmatrix} = \begin{pmatrix} f_1 + \frac{u_0(a_{1/2})}{h} \\ f_2 \\ \cdot \\ \cdot \\ f_{i_0-2} \\ f_{i_0-1} + \frac{u_{i_0}}{h^2} \end{pmatrix}.$$

We define the operator $P_{i_0} \in \mathcal{L}(\mathbb{R}^{p-1}, \mathbb{R}^{p-1})$ by:

$$P_{i_0}\gamma_{i_0} = \frac{\gamma_{i_0} - \gamma_{i_0-1}}{h} \quad \forall \gamma_{i_0} \in \mathbb{R}^{p-1},$$

and set:

$$\xi_{i-1/2} = \frac{u_i - u_{i-1}}{h} \in \mathbb{R}^{p-1} \;\; \forall\, i = 2, ..., N$$

and $\xi_{1/2} = -u_0(a_{1/2})$. We thus find that $\xi_{i-1/2} = P_i u_i + w_{i-1/2}$ for each $i \in \{2, ..., N\}$. The above relation also holds for $i = 1$ as, using the fact that u_1 is arbitrary, we obtain the initial conditions $P_1 = 0$ and $w_{1/2} = -u_0(a_{1/2})$. Therefore, we have:

$$\xi_{i+1/2} = P_{i+1}u_{i+1} + w_{i+1/2} \quad \forall i \in \{0, ..., N-1\}. \qquad [5.3]$$

However, as u_h is a solution of $(\mathcal{P}_h)$, we have:

$$u_1 - h^2\nabla^2_{h,2}u_1 - u_2 = h^2 f_1 + h u_0(a_{1/2}),$$

$$-u_{i-1} + 2u_i - h^2\nabla^2_{h,2}u_i - u_{i+1} = h^2 f_i \quad \forall i \in \{2, ..., N-1\},$$

which, together with the conditions $P_1 = 0$ and $w_{1/2} = -u_0(a_{1/2})$, implies that:

$$-\nabla^2_{h,2}u_i - f_i = P_{i+1}\frac{(u_{i+1} - u_i)}{h} + \frac{(P_{i+1} - P_i)}{h}u_i + \frac{(w_{i+1/2} - w_{i-1/2})}{h},$$
$$\forall\, i \in \{1, ..., N-1\},$$

which is a discrete approximation of [1.18].

LEMMA 5.1.– For $i = 2, \cdots, N$, the matrix P_i is positive definite and has the same eigenvectors as $-\nabla^2_{h,2}$. Furthermore, $I - hP_i$ is positive for all $i = 1, \cdots, N$.

PROOF.– From [5.3], we have that:

$$u_i = (I - hP_{i+1})u_{i+1} - hw_{i+1/2}, \quad \forall i \in \{0, ..., N-1\}.$$

Thus:

$$\begin{aligned} & - \nabla^2_{h,2}(I - hP_{i+1})u_{i+1} + \nabla^2_{h,2} h w_{i+1/2} - f_i \\ & = P_{i+1}\frac{u_{i+1}}{h} - \frac{P_i}{h}(I - hP_{i+1})u_{i+1} + P_i w_{i+1/2} \\ & + \frac{(w_{i+1/2} - w_{i-1/2})}{h}, \end{aligned} \quad [5.4]$$

for $i \in \{1, ..., N-1\}$ with "arbitrary" u_{i+1}. From the terms involving u_{i+1} in [5.4], we obtain the discrete Riccati equation for P:

$$\begin{cases} -\dfrac{P_{i+1} - P_i}{h}(I - hP_{i+1}) = P^2_{i+1} + \nabla^2_{h,2}(I - hP_{i+1}) \\ \qquad\qquad\qquad\qquad \forall i \in \{1, ..., N-1\}, \\ P_1 = 0. \end{cases} \quad [5.5]$$

We deduce that:

$$(I + hP_i - h^2\nabla^2_{h,2})P_{i+1} = P_i - h\nabla^2_{h,2}. \quad [5.6]$$

When $i = 1$, we obtain the following:

$$P_2 = -(I - h^2\nabla^2_{h,2})^{-1} h\nabla^2_{h,2}.$$

Observe that $-\nabla^2_{h,2}$ is a positive matrix and so $I - h^2\nabla^2_{h,2}$ is also positive. As these matrices share the same eigenvectors, it follows that P_2 is also positive with the same eigenvectors. We now show by induction that if P_i is positive and has the same eigenvectors as $-\nabla^2_{h,2}$, then so does P_{i+1}. From [5.6], we have that:

$$P_{i+1} = (I + hP_i - h^2\nabla^2_{h,2})^{-1}(P_i - h\nabla^2_{h,2}), \quad [5.7]$$

where $P_i - h\nabla^2_{h,2}$ and $(I + hP_i - h^2\nabla^2_{h,2})^{-1}$ are positive with the same eigenvectors, from which we deduce that P_{i+1} is positive with the same eigenvectors. Finally, from [5.7], we have that:

$$I - hP_{i+1} = (I + hP_i - h^2\nabla^2_{h,2})^{-1}, \quad [5.8]$$

which we just showed is positive. □

From the proof of lemma 5.1, $I - hP_i$ is positive and so the matrix $(I - hP_i)^{-1}$ is well defined. Therefore, from [5.5], we obtain that:

$$\begin{cases} -\dfrac{P_{i+1} - P_i}{h} = P^2_{i+1}(I - hP_{i+1})^{-1} + \nabla^2_{h,2} \;\; \forall i \in \{1, ..., N-1\} \\ P_1 = 0. \end{cases} \quad [5.9]$$

Additionally, from the independent terms in [5.4], we have that:

$$\frac{w_{i+1/2} - w_{i-1/2}}{h} = -(P_i - h\nabla^2_{h,2})w_{i+1/2} - f_i.$$

Then, from [5.7] and [5.8], we find that:

$$(P_i - h\nabla^2_{h,2}) = (I - hP_{i+1})^{-1}P_{i+1} \quad [5.10]$$

and, therefore:

$$\begin{cases} \dfrac{w_{i+1/2} - w_{i-1/2}}{h} = -(I - hP_{i+1})^{-1}P_{i+1}w_{i+1/2} - f_i \\ \qquad\qquad \forall i \in \{1, ..., N-1\}, \\ w_{1/2} = -u_0(a_{1/2}), \end{cases} \quad [5.11]$$

which may be rewritten as $w_{i-1/2} - L_i w_{i+1/2} = hf_i$, for each $i \in \{1, \cdots, N-1\}$ by setting $L_i = I + hP_{i+1}(I - hP_{i+1})^{-1} = (I - hP_{i+1})^{-1} = I + hP_i - h^2\nabla^2_{h,2}$. Now equation [5.3] is equivalent to:

$$\begin{cases} \dfrac{u_{i+1} - u_i}{h} = P_{i+1}u_{i+1} + w_{i+1/2} \;\; \forall i \in \{1, ..., N-1\} \\ u_N = u_1(a_N), \end{cases}$$

which may be rewritten as $-u_i + U_{i+1}u_{i+1} = hw_{i+1/2}$ for each $i \in \{1, ..., N-1\}$, by setting $U_i = I - hP_i$ for each $i \in \{2, ..., N\}$. We now rewrite these equations for u_h and w in the matrix form. We find:

$$\frac{1}{h}\begin{pmatrix} L_1 & & & \\ -I & L_2 & & 0 \\ & \ddots & \ddots & \\ & & \ddots & \\ 0 & & \ddots & \\ & & -I & L_{N-1} \end{pmatrix} \frac{1}{h}\begin{pmatrix} I & -U_2 & & & \\ & I & -U_3 & 0 & \\ & & \ddots & \ddots & \\ & & & \ddots & \\ 0 & & & I & -U_{N-1} \\ & & & & I \end{pmatrix} u_h = F_h. \qquad [5.12]$$

In the left-hand side of linear system [5.12], the first matrix is block lower triangular and the second is block upper triangular with units along the diagonal. Therefore, this expression is a block LU factorization of the block tridiagonal matrix A_h. We can check this by developing the product: for $i \in \{1, ..., N-2\}$, we have:

$$-L_iU_{i+1} = -(I + hP_{i+1}(I - hP_{i+1})^{-1})(I - hP_{i+1}) = -I.$$

Also, for $i \in \{2, ..., N-1\}$, we have:

$$U_i + L_i = I - hP_i + I + hP_i - h^2\nabla^2_{h,2} = 2I - h^2\nabla^2_{h,2} = B_i.$$

Finally, by [5.5], for the case $i = 1$, we have:

$$L_1 = I + hP_1 - h^2\nabla^2_{h,2} = I - h^2\nabla^2_{h,2} = B_1 - h^2\nabla^2_{h,2} = B_1.$$

Thus, we indeed recover the matrix A_h.

REMARK 5.1.– We have obtained the following discretized version of system [1.22–1.24]:

$$\begin{cases} -\dfrac{P_{i+1} - P_i}{h} = P^2_{i+1}(I - hP_{i+1})^{-1} + \nabla^2_{h,2} \ \forall i \in \{1, ..., N-1\}, \\ P_1 \equiv 0. \end{cases}$$

$$\begin{cases} \dfrac{w_{i+1/2} - w_{i-1/2}}{h} = -P_{i+1}(I - hP_{i+1})^{-1}w_{i+1/2} - f_i \ \forall i \in \{1, ..., N-1\}, \\ w_{1/2} = -u_0(a_{1/2}), \end{cases}$$

$$\begin{cases} \dfrac{u_{i+1} - u_i}{h} = P_{i+1}u_{i+1} + w_{i+1/2} \quad \forall i \in \{1, ..., N-1\}, \\ u_N = u_1(a_N). \end{cases}$$

The discrete equations for P_i and w_i are implicit, and the discrete equation for u_i is explicit. The equation for P_i is an implicit discretization of equation [1.22] with an error of order $O(h)$ from the term $(I - hP_{i+1})^{-1}$. These kinds of term are typical for discrete-time Riccati equations (see, for example, [ÅST 84]). In section 5.3, we will use the distinct operator P_i leading to an explicit equation for w_i and an implicit equation for u_i.

Using [5.10], the above system may be written in an equivalent form as follows:

$$\begin{cases} -\dfrac{P_{i+1} - P_i}{h} = P_{i+1}(P_i - h\nabla^2_{h,2}) + \nabla^2_{h,2} \ \forall i \in \{1, ..., N-1\}, \\ P_1 \equiv 0. \end{cases} \quad [5.13]$$

$$\begin{cases} \dfrac{w_{i+1/2} - w_{i-1/2}}{h} = -(P_i - h\nabla^2_{h,2})w_{i+1/2} - f_i \\ \qquad\qquad \forall i \in \{1, ..., N-1\}, \\ w_{1/2} = -u_0(a_{1/2}), \end{cases} \quad [5.14]$$

$$\begin{cases} \dfrac{u_{i+1} - u_i}{h} = P_{i+1}u_{i+1} + w_{i+1/2} \ \ \forall i \in \{1, ..., N-1\}, \\ u_N = u_1(a_N). \end{cases} \quad [5.15]$$

Note that this form of P_{i+1} is calculated as a function of P_i with a linear relation at each stage i, and that, once P_i is known, it is not necessary to invert any matrix to calculate w_i. Additionally, [5.13], [5.14] and [5.15] are discretized $O(h)$ approximations of [1.22], [1.23] and [1.24], respectively. However, as mentioned earlier, u_h is an $O(h^2)$ approximation of u. We can show that the $O(h)$ errors in the calculation of $w_{i+1/2}$ and u_i mutually compensate.

REMARK 5.2.– In finite dimensions, after discretization, we can make explicit the singularity in the initial condition on P that was observed in Chapter 3 for certain boundary conditions. Consider the situation of a Dirichlet condition on the face Γ_0, $u|_{\Gamma_0} = u^0$, which is then discretized as $\dfrac{u_0 + u_1}{2} = u^0$ according to the chosen grid. Then, writing out equation [5.3] for $i = 0$ yields:

$$2\frac{u_1 - u^0}{h} = P_1 u_1 + w_{1/2},$$

from which we deduce that:

$$P_1 = 2\frac{I}{h} \qquad w_{1/2} = -2\frac{u^0}{h}.$$

The singularity associated with the Dirichlet condition at $x = 0$ is therefore $O(h^{-1})$. Numerical method [5.13] achieves stable numerical integration despite this initial condition. But this is not always the case. Imagine, for example, that we wish to discretize the continuous Riccati equation using an explicit $O(h)$ Euler method. Formula [5.13] becomes:

$$\begin{cases} -\dfrac{P_{i+1} - P_i}{h} = P_i^2 + \nabla_{h,2}^2 \ \forall i \in \{1, ..., N-1\}, \\ P_1 \equiv 0. \end{cases} \qquad [5.16]$$

It can be shown that with this initial condition, for h sufficiently small, the matrices P_i retain the property of positivity. However, this no longer holds in the case considered in this remark, where $P_1 = 2\frac{I}{h}$: already in the second step, the singularity is $0(h^{-2})$. We observe that decreasing h does not allow positivity to be retained, and so, this method is unstable.

REMARK 5.3.– The backward factorization calculation in section 1.3, after discretization, naturally leads to the product of a block upper triangular matrix and a block lower triangular matrix, such as that presented in section A.2.1 of the Appendix.

5.3. A second method of discretization

Instead of defining the operator P_i as in section 5.2, we can define the operator $P_{i_0} \in \mathcal{L}(\mathbb{R}^{p-1}, \mathbb{R}^{p-1})$ by:

$$P_{i_0}\gamma_{i_0} = \frac{\gamma_{i_0+1} - \gamma_{i_0}}{h} \quad \forall \gamma_{i_0} \in \mathbb{R}^{p-1}.$$

Here, the components $\gamma_1, ..., \gamma_{i_0-1}, \gamma_{i_0+1}$ are calculated from the data γ_{i_0}. We need to show that this is possible. Indeed, γ_{i_0+1} is the (i_0+1)-th component of the vector:

$$\gamma = \begin{pmatrix} \gamma_1 \\ \cdot \\ \cdot \\ \gamma_{i_0+1} \end{pmatrix} \in \mathbb{R}^{(i_0+1)(p-1)},$$

which satisfies:

$$\frac{1}{h^2}\begin{pmatrix} B_1 & -I & & & & \\ -I & B_2 & -I & & 0 & \\ & . & . & . & & \\ & & . & . & . & \\ & 0 & & -I & . & -I \\ & & & & -I & B_{i_0} \end{pmatrix}\begin{pmatrix} \gamma_1 \\ . \\ . \\ . \\ \gamma_{i_0-1} \\ \gamma_{i_0} \end{pmatrix} = \begin{pmatrix} 0 \\ . \\ . \\ . \\ 0 \\ \frac{\gamma_{i_0+1}}{h^2} \end{pmatrix}.$$

This is equivalent to:

$$\frac{1}{h^2}\begin{pmatrix} B_1 & -I & & & & \\ -I & B_2 & -I & & 0 & \\ & . & . & . & & \\ & & . & . & -I & \\ & 0 & & -I & B_{i_0-1} & 0 \\ & & & & -I & -I \end{pmatrix}\begin{pmatrix} \gamma_1 \\ . \\ . \\ . \\ \gamma_{i_0-1} \\ \gamma_{i_0+1} \end{pmatrix} = \begin{pmatrix} 0 \\ . \\ . \\ 0 \\ \frac{\gamma_{i_0}}{h^2} \\ \frac{-B_{i_0}\gamma_{i_0}}{h^2} \end{pmatrix},$$

for $i_0 \in \{2, ..., N-1\}$, and is equivalent to $\gamma_2 = B_1\gamma_1$ for $i_0 = 1$. For each $i_0 \in \{1, ..., N-1\}$, we can also define the vector $w_{i_0+1/2} \in \mathbb{R}^{p-1}$ by $w_{i_0+1/2} = \frac{\beta_{i_0+1}}{h}$, where β_{i_0+1} is the (i_0+1)-th component of the vector:

$$\beta = \begin{pmatrix} \beta_1 \\ . \\ . \\ \beta_{i_0-1} \\ 0 \\ \beta_{i_0+1} \end{pmatrix} \in \mathbb{R}^{(i_0+1)(p-1)}.$$

This vector satisfies:

$$\frac{1}{h^2}\begin{pmatrix} B_1 & -I & & & & \\ -I & B_2 & -I & & 0 & \\ & . & . & . & & \\ & & . & . & . & \\ & 0 & & -I & B_{i_0-1} & -I \\ & & & & -I & B_{i_0} \end{pmatrix}\begin{pmatrix} \beta_1 \\ . \\ . \\ . \\ \beta_{i_0-1} \\ 0 \end{pmatrix} = \begin{pmatrix} f_1 + \frac{u_0(a_{1/2})}{h} \\ f_2 \\ . \\ . \\ f_{i_0-1} \\ f_{i_0} + \frac{\beta_{i_0+1}}{h^2} \end{pmatrix},$$

for $i_0 \in \{2, ..., N-1\}$, which is equivalent to:

$$\frac{1}{h^2}\begin{pmatrix} B_1 & -I & & & & \\ -I & B_2 & -I & & 0 & \\ & \cdot & \cdot & \cdot & & \\ & & \cdot & \cdot & -I & \\ & 0 & & -I & B_{i_0-1} & 0 \\ & & & & -I & -I \end{pmatrix}\begin{pmatrix} \beta_1 \\ \cdot \\ \cdot \\ \cdot \\ \beta_{i_0-1} \\ \beta_{i_0+1} \end{pmatrix} = \begin{pmatrix} f_1 + \frac{u_0(a_{1/2})}{h} \\ f_2 \\ \cdot \\ \cdot \\ f_{i_0-1} \\ f_{i_0} \end{pmatrix},$$

and satisfies $\frac{1}{h^2}B_1 0 = f_1 + \frac{u_0(a_{1/2})}{h} + \frac{\beta_2}{h^2}$ for $i_0 = 1$, which implies that $\beta_2 = -h^2 F_1$. Here, we explicitly observe that the vector $\gamma + \beta$ satisfies:

$$\frac{1}{h^2}\begin{pmatrix} B_1 & -I & & & & \\ -I & B_2 & -I & & 0 & \\ & \cdot & \cdot & \cdot & & \\ & & \cdot & \cdot & \cdot & \\ & 0 & & -I & B_{i_0-1} & 0 \\ & & & & -I & -I \end{pmatrix}\begin{pmatrix} \gamma_1+\beta_1 \\ \cdot \\ \cdot \\ \cdot \\ \gamma_{i_0-1}+\beta_{i_0-1} \\ \gamma_{i_0+1}+\beta_{i_0+1} \end{pmatrix} = \begin{pmatrix} f_1+\frac{u_0(a_{1/2})}{h} \\ f_2 \\ \cdot \\ \cdot \\ f_{i_0-1}+\frac{\gamma_{i_0}}{h^2} \\ f_{i_0} - \frac{B_{i_0}\gamma_{i_0}}{h^2} \end{pmatrix}.$$

We therefore find that:

$$\xi_{i+1/2} = P_i u_i + w_{i+1/2} \quad \forall i \in \{1, ..., N-1\}.$$

Then, setting $L_i = (I - hP_i(I+hP_i)^{-1}) = (I+hP_i)^{-1}$ and $U_i = I + hP_i$, and adopting an approach similar to that of section 5.2, we reach the following block LU factorization of problem $(\mathcal{P}_h)$:

$$\frac{1}{h}\begin{pmatrix} I & & & & \\ -L_1 & I & & 0 & \\ & \cdot & \cdot & & \\ & & \cdot & \cdot & \\ & 0 & \cdot & \cdot & \\ & & & -L_{N-2} & I \end{pmatrix}\frac{1}{h}\begin{pmatrix} U_1 & -I & & & \\ & \cdot & -I & 0 & \\ & & \cdot & \cdot & \\ & & & \cdot & \cdot \\ & 0 & & \cdot & -I \\ & & & & U_{N-1} \end{pmatrix} u_h = F_h.$$

[5.17]

Now, the discretized version of system [1.22–1.24] thus obtained is:

$$\begin{cases} -\dfrac{P_i - P_{i-1}}{h} = P_{i-1}(I + hP_{i-1})^{-1}P_{i-1} + \nabla^2_{h,2} \\ \qquad\qquad \forall i \in \{1, ..., N-1\}, \\ P_1 = 0. \end{cases} \qquad [5.18]$$

$$\begin{cases} \dfrac{w_{i+1/2} - w_{i-1/2}}{h} = -P_{i-1}(I + hP_{i-1})^{-1}w_{i-1/2} - f_i \\ \qquad\qquad \forall i \in \{1, ..., N-1\}, \\ w_{1/2} = -u_0(a_{1/2}), \end{cases} \qquad [5.19]$$

$$\begin{cases} \dfrac{u_{i+1} - u_i}{h} = P_i u_i + w_{i+1/2} \;\; \forall i \in \{1, ..., N-1\}, \\ u_N = u_1(a_N), \end{cases} \qquad [5.20]$$

Unlike section 5.2, the equation for w_i is explicit and the equation for u_i is implicit. The equation for P_i still requires the inverse to be calculated. The two factorizations [5.12] and [5.17] correspond to two situations with units along the main diagonal in either the upper triangular factor or the lower triangular factor.

To gain insight into the relation between these two factorizations, denote P_i' as the operator defined in section 5.2 by [5.9] and P_i the operator defined in this section by [5.18]. We can check that they are related by:

$$(I + hP_i)^{-1} = I - hP_i' \;\; \forall i.$$

5.4. A third possibility: centered scheme

Another possible way to define P_i is given by the following $O(h^2)$ centered finite differences method:

$$\frac{u_{i+1} - u_i}{h} = P_i\left(\frac{u_i + u_{i+1}}{2}\right) + w_{i+1/2} \;\; \forall i \in \{1, ..., N-1\},$$
$$u_N = u_1(a_N). \qquad [5.21]$$

This gives a Crank–Nicolson-type discretization of the equation for u. By the same approach as in the previous sections, we form the finite differences

$\frac{1}{h}(u_{i+1} - u_i)$ and $\frac{1}{h}(u_i - u_{i-1})$, which we express as a function of u_i by using [5.21]. This yields:

$$\begin{aligned} u_{i+1} - 2u_i + u_{i-1} = &\Big[\frac{hP_i}{2}(I + (I - \frac{hP_i}{2})^{-1}(I + \frac{hP_i}{2}) \\ &- \frac{hP_{i-1}}{2}(I + (I + \frac{hP_{i-1}}{2})^{-1}(I - \frac{hP_{i-1}}{2})\Big] u_i \\ &+ h(I + \frac{hP_i}{2})(I - \frac{hP_i}{2})^{-1} w_{i+1/2} \\ &- h(I - \frac{hP_{i-1}}{2})(I + \frac{hP_{i-1}}{2})^{-1} w_{i-1/2} \\ = &- h^2 \nabla^2_{h,2} - h^2 f_i. \end{aligned} \quad [5.22]$$

By separating the component depending on u_i, which is arbitrary, from the independent component, we obtain the following equation satisfied by P_i:

$$\begin{cases} -\frac{P_{i+1} - P_i}{h} = \frac{1}{2} P_i (I + \frac{hP_i}{2})^{-1} P_i + \frac{1}{2} P_{i+1}(I - \frac{hP_{i+1}}{2})^{-1} P_{i+1} \\ \qquad + \nabla^2_{h,2}, \\ P_1 = 0, \end{cases} \quad [5.23]$$

and the equation satisfied by w_i:

$$\begin{cases} \frac{w_{i+1/2} - w_{i-1/2}}{h} + \frac{1}{2} P_i (I - \frac{hP_i}{2})^{-1} w_{i+1/2} \\ \qquad + \frac{1}{2} P_{i-1}(I + \frac{hP_{i-1}}{2})^{-1} w_{i-1/2} = -f_i, \\ w_{1/2} = -u_0(a_{1/2}), \end{cases} \quad [5.24]$$

which, together with [5.21], completes the factorization of the discretized system. Note that here, the approximations of the derivatives w.r.t. x are centered finite differences and so the approximations are $O(h^2)$ in each of the three equations.

It is not necessary to solve a matrix equation to obtain P_{i+1} as a function of P_i, as formula [5.23] seems to imply. After a series of algebraic manipulations, formula [5.23] can be rewritten as:

$$\frac{P_i}{h} = \frac{1}{2} P_i (I + \frac{h}{2} P_i)^{-1} P_i + \frac{1}{h}(I - \frac{h}{2} P_{i+1})^{-1} P_{i+1} + \nabla^2_{h,2}. \quad [5.25]$$

After left-multiplying [5.25] by $(I-\frac{h}{2}P_{i+1})$ and regrouping the terms with P_{i+1}, we find:

$$P_{i+1}(I+\frac{h}{2}P_i-\frac{h^2}{4}P_i(I+\frac{h}{2}P_i)^{-1}P_i-\frac{h^2}{2}\nabla^2_{h,2}) = P_i(I-\frac{h}{2}(I+\frac{h}{2}P_i)^{-1}P_i) - h\nabla^2_{h,2}. \quad [5.26]$$

We deduce that:

$$P_{i+1} = (I+\frac{h}{2}P_i(I+\frac{h}{2}P_i)^{-1}-\frac{h^2}{2}\nabla^2_{h,2})^{-1}(P_i(I+\frac{h}{2}P_i)^{-1}-h\nabla^2_{h,2}). \quad [5.27]$$

Thus, P_{i+1} may be obtained from P_i by two inversions and two matrix multiplications.

We now wish to show that this calculation may be interpreted as a classical example of factorization in linear algebra. We introduce the change of variables:

$$\overline{w}_{i+1/2} = (I-\frac{hP_i}{2})^{-1/2}(I+\frac{hP_i}{2})^{-1/2}w_{i+1/2}.$$

Then, u_i satisfies:

$$\begin{cases} (I-\frac{hP_i}{2})^{1/2}(I+\frac{hP_i}{2})^{-1/2}u_{i+1} \\ \qquad -(I-\frac{hP_i}{2})^{-1/2}(I+\frac{hP_i}{2})^{1/2}u_i = h\overline{w}_{i+1/2}, \\ u_N = u_1(a_N), \end{cases}$$

and the equation for $\overline{w}_{i+1/2}$ is:

$$\begin{cases} (I-\frac{hP_i}{2})^{-1/2}(I+\frac{hP_i}{2})^{1/2}\overline{w}_{i+1/2} \\ \qquad -(I-\frac{hP_{i-1}}{2})^{1/2}(I+\frac{hP_{i-1}}{2})^{-1/2}\overline{w}_{i-1/2} = -hf_i, \\ \overline{w}_{1/2} = -u_0(a_{1/2}). \end{cases}$$

The factorization can now be written as:

$$h^2 A_h = \begin{pmatrix} C_1 & & & & \\ -D_1 & C_2 & & 0 & \\ & \ddots & \ddots & & \\ & & \ddots & \ddots & \\ & 0 & & \ddots & \ddots \\ & & & -D_{N-2} & C_{N-1} \end{pmatrix} \begin{pmatrix} C_1 & -D_1 & & & \\ & C_2 & \ddots & 0 & \\ & & \ddots & \ddots & \\ & & & \ddots & \ddots \\ & 0 & & \ddots & -D_{N-2} \\ & & & & C_{N-1} \end{pmatrix}, \quad [5.28]$$

where

$$C_i = (I - \frac{hP_i}{2})^{-1/2}(I + \frac{hP_i}{2})^{1/2},$$
$$D_i = (I + \frac{hP_i}{2})^{-1/2}(I - \frac{hP_i}{2})^{1/2}.$$

This factorization corresponds to a block Cholesky factorization of A_h. This shows the connection between the well-established properties of the precision and the stability of the centered finite difference approximations (Crank–Nicolson method) and the stability of the Cholesky factorization.

Also, note that taking the square root (or, more precisely, taking the $1/2$-th power), which is an inherent step of the Cholesky method, appears in formulation [5.28]. However, it is not numerically necessary as it does not appear in the equivalent formulation [5.12], [5.23–5.24].

5.5. Row permutation

We now wish to consider the effect of block row permutation and show the connection with the continuous calculation in section 3.5. We adopt the conditions of the calculation in section 5.2. Consider the matrix A_h of problem $(\mathcal{P}_h)$. It is a well-known fact that the Gaussian method with row permutation is theoretically equivalent to a method that first performs all permutations and then performs LU factorization on the permuted matrix. In general, this is only relevant in a theoretical setting because at the k-th step the decision to perform a permutation is made based on the state of the factorization calculation at that step. However, in our case, we have block tridiagonal matrices and there is therefore only one single possible block row permutation at each step: the permutation that moves the sub-diagonal block into the diagonal position. Suppose we decide to perform this permutation systematically. In this case, the first row block ends up in last position, whereas all other blocks "move up" by one step:

$$A'_h = \frac{1}{h^2}\begin{pmatrix} -I & B_2 & -I & & 0 & \\ & & \cdot & \cdot & \cdot & \\ & & & \cdot & \cdot & \cdot \\ & 0 & & -I & \cdot & -I \\ & & & & -I & B_{N-1} \\ B_1 & -I & & & 0 & 0 \end{pmatrix}. \quad [5.29]$$

The problem that we must solve then becomes:

$$A'_h u_h = \begin{pmatrix} f_2 \\ \cdot \\ f_{N-2} \\ f_{N-1} + \frac{u_1(a_N)}{h^2} \\ f_1 + \frac{u_0(a_{1/2})}{h} \end{pmatrix}. \quad [5.30]$$

We apply the method of invariant embedding as in section 5.2. At step $i_0 \in \{2, \ldots, N\}$, we have:

$$\frac{1}{h^2}\begin{pmatrix} -I & B_2 & -I & & 0 & \\ & \cdot & \cdot & \cdot & & \\ & & \cdot & \cdot & \cdot & \\ & 0 & & -I & \cdot & -I \\ & & & & -I & B_{i_0-1} \\ & & & & & -I \end{pmatrix}\begin{pmatrix} u_1 \\ \cdot \\ \cdot \\ \cdot \\ \cdot \\ u_{i_0-1} \end{pmatrix} = \begin{pmatrix} f_2 \\ \cdot \\ \cdot \\ f_{i_0-2} \\ f_{i_0-1} + \frac{\gamma_{i_0}}{h^2} \\ f_{i_0} + \frac{\gamma_{i_0+1}}{h^2} - \frac{1}{h^2} B_{i_0}\gamma_{i_0} \end{pmatrix}.$$

The solution has an affine dependency on the data γ_{i_0} and γ_{i_0+1}. As the system is upper triangular with $-I$ on the diagonal, this solution can be obtained by simple successive substitution. However, as we want the solution u_h to satisfy the condition $\dfrac{u_1 - u_0}{h} = -u_0(a_{1/2})$, the data cannot be independent but must satisfy an affine relation. We write this relation in the form:

$$\frac{\gamma_{i_0+1} - \gamma_{i_0}}{h} = P_{i_0+1}\gamma_{i_0+1} + w_{i_0+1/2}. \qquad [5.31]$$

We check that P_i and $w_{i+1/2}$ are, respectively, solutions of [5.11] and [5.9]. For $i_0 = 0$, regardless of the value of the data γ_1, we want $\dfrac{\gamma_1 - \gamma_0}{h} = -u_0(a_{1/2})$. We therefore deduce that:

$$P_1 = 0, \quad w_{1/2} = -u_0(a_{1/2}). \qquad [5.32]$$

Consider the solution of [5.30] $\{u_j\}_{j=1}^{i_0-1}$ corresponding to the data γ_{i_0}, γ_{i_0+1}. We have:

$$\frac{\gamma_{i_0} - u_{i_0-1}(\gamma_{i_0}, \gamma_{i_0+1})}{h} = P_{i_0}\gamma_{i_0} + w_{i_0-1/2}, \qquad [5.33]$$

because $\{u_j\}_{j=1}^{i_0-2}$ is a solution of the same problem of dimension $(i_0-2)(p-1)$ for the data $u_{i_0-1}(\gamma_{i_0}, \gamma_{i_0+1})$ and γ_{i_0}. By subtracting [5.33] from [5.31]:

$$\begin{aligned}\frac{\gamma_{i_0+1} - 2\gamma_{i_0} + u_{i_0-1}}{h^2} &= -f_{i_0} - \nabla^2_{h,2}\gamma_{i_0}\\ &= P_{i_0}\frac{\gamma_{i_0+1} - \gamma_{i_0}}{h} + \frac{P_{i_0+1} - P_{i_0}}{h}\gamma_{i_0}\\ &\quad + \frac{(w_{i_0+1/2} - w_{i_0-1/2})}{h}.\end{aligned} \qquad [5.34]$$

By applying [5.31] to reduce to just γ_{i_0}:

$$\frac{\gamma_{i_0+1} - \gamma_{i_0}}{h} = (I - hP_{i_0+1})^{-1}(P_{i_0+1}\gamma_{i_0} + w_{i_0+1/2}). \qquad [5.35]$$

By substituting this into [5.34] and using the fact that γ_{i_0} is arbitrary, we have that:

$$-\frac{P_{i_0+1} - P_{i_0}}{h} = P_{i_0+1}(I - hP_{i_0+1})^{-1}P_{i_0+1} + \nabla^2_{h,2}, \qquad [5.36]$$

and for the independent term γ_{i_0}, we have:

$$\frac{w_{i_0+1/2} - w_{i_0-1/2}}{h} = -P_{i_0+1}(I - hP_{i_0+1})^{-1}w_{i_0+1/2} - f_{i_0}. \qquad [5.37]$$

These equations are satisfied for i_0 ranging from 2 to $N-1$. From the initial conditions [5.32], we also recover [5.11] and [5.9]. However, the "upward" phase for calculating u_i is now accomplished by solving [5.30] by substitution, which we can write as:

$$-u_{i-1} + 2u_i - h^2\nabla^2_{h,2}u_i - u_{i+1} = h^2 f_i \quad \forall i \in \{2, ..., N-1\}, \qquad [5.38]$$

adding the following equation to find u_{N-1}:

$$\frac{u_1(a_N) - u_{N-1}}{h} = P_N u_1(a_N) + w_{N-1/2}. \qquad [5.39]$$

This calculation shows that the block LU factorization with systematic block row permutation may be interpreted as a discretization of the method of invariant embedding in the form of a family of Cauchy problems that was presented in section 3.5.

REMARK 5.4.– It is a well-known fact that the stability of the Gaussian method is improved by performing the row permutation that brings the sub-diagonal element with the highest absolute value into the pivotal position. When discretizing the second derivatives, as is the case here, the blocks with the highest spectral radius naturally end up in the diagonal position: indeed, in our case, if $\lambda_{h,i}$ are the eigenvalues of $-\nabla^2_{h,2}$, the diagonal term of A_h presented in [5.1] has the eigenvalues $2(1 + h\lambda_{h,i})$, whereas the sub-diagonal blocks have the eigenvalue -1.

There is therefore no benefit in performing block row permutations from the point of view of stability. The above calculation even shows that, in our case, the problem with row permutations leads to an ill-posed problem in the limit $h \longrightarrow 0$: the Cauchy problem for the Laplacian. This shows the connection between the destabilization of the Gaussian factorization with pivoting and the instability of the Cauchy problem. However, as commented in section 3.5, in the formulation corresponding to the Cauchy problem, storing the matrices P_i is not necessary to calculate u_i.

5.6. Case of a discretization of the section by finite elements

The principle of factorizing the discretized problem by invariant embedding works in general, as explained in the Appendix. To show that this method does not depend on the discretization by finite differences, we will give an example here where the section is discretized by finite elements.

Suppose again that we are in the case of the cylinder $\Omega =]0,a[\times\mathcal{O}$ with the discretization in x from the beginning of this chapter, with $x_i = (-1/2+i)h$, $i \in \{0,...,N\}$. Then, if $u \in \mathcal{C}^3(\Omega)$ with $\Omega =]0,a[\times\mathcal{O})$, $i \in \{2,...,N-1\}$ and $y \in \mathcal{O}$:

$$-\Delta u(x_i,y) = \frac{1}{h^2}\left(-u(x_{i-1},y)+2u(x_i,y)-u(x_{i+1},y)\right)-\Delta_y u(x_i,y)+O(h)$$

and if $i=1$:

$$-\Delta u(x_1,y) = \frac{1}{h^2}\left(u(x_1,y)-u(x_2,y)\right)-\Delta_y u(x_1,y)-\frac{1}{h}u_0(y)+O(h).$$

Note here that for $i = N$ (corresponding to the section Γ_a), we have $u(x_N,y) = u_1(y)$ and for $i \in \{1,...,N-1\}$ and $y \in \partial\mathcal{O}$, $u(x_i,y) = 0$ (lateral boundary Σ).

Then, by a semi-discretization in x, $u(x_i,y)$, $i \in \{1,...,N\}$ can be approached by the solution $u_i(y)$ of the family of coupled elliptic problems:

$$\frac{1}{h^2}\left(u_1(y)-u_2(y)\right)-\Delta_y u_1(y) = f(x_1,y)+\frac{1}{h}u_0(y), \quad y \in \mathcal{O}, \tag{5.40}$$

$$\frac{1}{h^2}\left(-u_{i-1}(y)+2u_i(y)-u_{i+1}(y)\right)-\Delta_y u_i(y) = f(x_i,y), \quad y \in \mathcal{O}, \quad i \in \{2,...,N\}. \tag{5.41}$$

The weak formulation of system [5.40–5.41] is:

$$\int_{\mathcal{O}} \frac{1}{h^2}(u_1(y) - u_2(y))\,\varphi(y)\mathrm{d}y + \int_{\mathcal{O}} \nabla_y u_1(y)\nabla_y\varphi(y)\mathrm{d}y$$
$$= \int_{\mathcal{O}} \left(f(x_1, y) + \frac{1}{h}u_0(y)\right)\varphi(y)\mathrm{d}y, \quad \forall\varphi \in H_0^1(\mathcal{O}). \qquad [5.42]$$

$$\int_{\mathcal{O}} \frac{1}{h^2}\left(-u_{i-1}(y) + 2u_i(y) - u_{i+1}(y)\right)\varphi(y)\mathrm{d}y$$
$$+ \int_{\mathcal{O}} \nabla_y u_i(y)\nabla_y\varphi(y)\mathrm{d}y = \int_{\mathcal{O}} f(x_i, y)\varphi(y)\mathrm{d}y, \quad y \in \mathcal{O},$$
$$\forall\varphi \in H_0^1(\mathcal{O}). \qquad [5.43]$$

In the case where the Neumann data u_0 is less regular and defined in $H_{00}^{1/2}(\mathcal{O})'$, formula [5.42] still holds by replacing $\int_{\mathcal{O}} u_0(y)\varphi(y)\,\mathrm{d}y$ with the duality pairing $< u_0, \varphi >_{H_{00}^{1/2}(\mathcal{O})' \times H_{00}^{1/2}(\mathcal{O})}$. We will use the same notation in the following for decompositions over a basis of functions that will be determined from a triangulation.

Let $\mathcal{T}_r$ be a triangulation of $\mathcal{O}$ with r the largest diameter of the circumcircles of the triangles. Let k be the number of interior vertices and V_r the vector space generated by functions of the form $\{\varphi_1, \cdots, \varphi_k\}$ that are P^1 with respect to the interior nodes. This space provides an approximation of $H_0^1(\mathcal{O})$. Thus, if $M = \left(\int_{\mathcal{O}} \varphi_i(y)\varphi_j(y)\mathrm{d}y\right)_{i,j=1}^k$ is the mass matrix, $K = \left(\int_{\mathcal{O}} \nabla_y\varphi_i(y)\nabla_y\varphi_j(y)\mathrm{d}y\right)_{i,j=1}^k$ is the stiffness matrix, $B_1 = M + h^2K$, and $B_i = 2M + h^2K$ for $i = 2, ..., N-1$, we have (see [RAM 12]): $u(x_i, y) \simeq u_i(y) \simeq \sum_{j=1}^k u_i^j\varphi_j(y)$, with:

$$u_{h,r} = \begin{pmatrix} u_1 \\ u_2 \\ \cdot \\ \cdot \\ \\ u_{N-1} \end{pmatrix} = \begin{pmatrix} u_1^1 \\ \cdot \\ \cdot \\ u_1^k \\ u_2^1 \\ \cdot \\ u_{N-1}^k \end{pmatrix} \in \mathbb{R}^{(N-1)k}$$

satisfying:

$$(\mathcal{P}_{h,r}) \qquad A_{h,r} u_{h,r} = F_{h,r},$$

where:

$$A_{h,r} = \frac{1}{h^2} \begin{pmatrix} B_1 & -M & & & & \\ -M & B_2 & -M & & 0 & \\ & \cdot & \cdot & \cdot & & \\ & & \cdot & \cdot & \cdot & \\ & 0 & & -M & \cdot & -M \\ & & & & -M & B_{N-1} \end{pmatrix} \in \mathcal{M}_{(N-1)k \times (N-1)k}(\mathbb{R}),$$

$$F_{h,r} = \begin{pmatrix} F_1 \\ \cdot \\ \cdot \\ \cdot \\ F_{N-1} \end{pmatrix} = \begin{pmatrix} f_1 + \left(\int_{\mathcal{O}} \frac{1}{h} u_0(y) \varphi_j(y) \mathrm{d}y\right)_{j=1}^k \\ f_2 \\ \cdot \\ \cdot \\ \cdot \\ f_{N-2} \\ f_{N-1} + \left(\int_{\mathcal{O}} \frac{u_1(y)}{h^2} \varphi_j(y) \mathrm{d}y\right)_{j=1}^k \end{pmatrix},$$

and $f_i = \left(\int_{\mathcal{O}} f(x_i, y) \varphi_j(y) \mathrm{d}y\right)_{j=1}^k \in \mathbb{R}^k$, for all $i \in \{1, ..., N-1\}$.

Following the same procedure as in section 5.2, we set:

$$\xi_{i-1/2} = \frac{u_i - u_{i-1}}{h} \in \mathbb{R}^k \;\; \forall\, i = 2, ..., N$$

and $\xi_{1/2} = -u_{0,r}$, with $u_{0,r}$ the coordinates in V_r of an approximation of u_0. For each $i_0 \in \{2, ..., N\}$, we define a problem similar to $(\mathcal{P}_{h,r})$ on the subdomain $1 \leq i \leq i_0$ with the Dirichlet data u_{i_0} at $i = i_0$. Let γ_i be the component of u_i that depends linearly on the data $\gamma_{i_0} = u_{i_0}$; we define the operator $P_{i_0} \in \mathcal{L}(\mathbb{R}^k, \mathbb{R}^k)$ by:

$$P_{i_0} \gamma_{i_0} = \frac{\gamma_{i_0} - \gamma_{i_0-1}}{h} \quad \forall \gamma_{i_0} \in \mathbb{R}^k,$$

where γ_{i_0-1} is the $(i_0 - 1)$-th component of the vector:

$$\gamma = \begin{pmatrix} \gamma_1 \\ \cdot \\ \cdot \\ \gamma_{i_0-1} \end{pmatrix} \in \mathbb{R}^{(i_0-1)k},$$

satisfying the following problem homogeneous with respect to the data γ_{i_0}:

$$\frac{1}{h^2}\begin{pmatrix} B_1 & -M & & & & \\ -M & B_2 & -M & & 0 & \\ & \cdot & \cdot & \cdot & & \\ & & \cdot & \cdot & \cdot & \\ & 0 & & -M & \cdot & -M \\ & & & & -M & B_{i_0-1} \end{pmatrix}\begin{pmatrix} \gamma_1 \\ \cdot \\ \cdot \\ \cdot \\ \gamma_{i_0-2} \\ \gamma_{i_0-1} \end{pmatrix} = \begin{pmatrix} 0 \\ \cdot \\ \cdot \\ \cdot \\ 0 \\ M\frac{\gamma_{i_0}}{h^2} \end{pmatrix}. \quad [5.44]$$

For each $i_0 \in \{2, ..., N\}$, we also define the vector $w_{i_0-1/2} \in \mathbb{R}^k$ by $w_{i_0-1/2} = \frac{-\beta_{i_0-1}}{h}$, where β_{i_0-1} is the (i_0-1)-th component of the vector:

$$\beta = \begin{pmatrix} \beta_1 \\ \cdot \\ \cdot \\ \beta_{i_0-1} \end{pmatrix} \in \mathbb{R}^{(i_0-1)k}$$

satisfying:

$$\frac{1}{h^2}\begin{pmatrix} B_1 & -M & & & & \\ -M & B_2 & -M & & 0 & \\ & \cdot & \cdot & \cdot & & \\ & & \cdot & \cdot & \cdot & \\ & 0 & & -M & \cdot & -M \\ & & & & -M & B_{i_0-1} \end{pmatrix}\begin{pmatrix} \beta_1 \\ \cdot \\ \cdot \\ \cdot \\ \beta_{i_0-2} \\ \beta_{i_0-1} \end{pmatrix} =$$

$$\begin{pmatrix} f_1 + \left(\int_{\mathcal{O}} \frac{1}{h} u_0(y)\varphi_j(y)\mathrm{d}y\right)_{j=1}^k \\ f_2 \\ \cdot \\ \cdot \\ f_{i_0-2} \\ f_{i_0-1} \end{pmatrix}.$$

Then, the vector $u = \gamma + \beta$ satisfies:

$$\frac{1}{h^2}\begin{pmatrix} B_1 & -M & & & & \\ -M & B_2 & -M & & 0 & \\ & \cdot & \cdot & \cdot & & \\ & & \cdot & \cdot & \cdot & \\ & 0 & & -M & \cdot & -M \\ & & & & -M & B_{i_0-1} \end{pmatrix}\begin{pmatrix} u_1 \\ \cdot \\ \cdot \\ \cdot \\ \cdot \\ u_{i_0-1} \end{pmatrix} =$$

$$\begin{pmatrix} f_1 + \frac{1}{h}\left(\int_{\mathcal{O}} u_0(y)\varphi_j(y)\mathrm{d}y\right)_{j=1}^k \\ f_2 \\ \cdot \\ \cdot \\ f_{i_0-2} \\ f_{i_0-1} + M\frac{u_{i_0}}{h^2} \end{pmatrix}.$$

We thus obtain that $\xi_{i-1/2} = P_i u_i + w_{i-1/2}$ for each $i \in \{2, ..., N\}$. This relation also holds for $i = 1$ because, using the fact that u_1 is arbitrary, we obtain the initial conditions $P_1 = 0$ and $w_{1/2} = -u_{0,r}$. We therefore have:

$$\xi_{i+1/2} = P_{i+1}u_{i+1} + w_{i+1/2} \quad \forall i \in \{0, ..., N-1\}. \qquad [5.45]$$

PROPOSITION 5.1.– The matrices MP_i are symmetric and positive definite for all $i = 2, \cdots, N$.

PROOF.– Let $i_0 \in \{2, \cdots, N\}$ and $\gamma_{i_0} \in \mathbb{R}^k$ be arbitrary. By left-multiplying [5.44] by $h^2\gamma^T$, we find that:

$$\gamma_1^T B_1\gamma_1 - \gamma_1^T M\gamma_2 + \sum_{i=2}^{i_0-2}\left(-\gamma_i^T M\gamma_{i-1} + \gamma_i^T B_i\gamma_i - \gamma_i^T M\gamma_{i+1}\right)$$
$$+ \left(-\gamma_{i_0-1}^T M\gamma_{i_0-2} + \gamma_{i_0-1}^T B_i\gamma_{i_0-1}\right) = \gamma_{i_0-1}^T M\gamma_{i_0},$$

which by definition of the B_i is:

$$\gamma_1^T M\gamma_1 + h^2\gamma_1^T K\gamma_1 - \gamma_1^T M\gamma_2$$
$$+ \sum_{i=2}^{i_0-2}\left(-\gamma_i^T M\gamma_{i-1} + 2\gamma_i^T M\gamma_i + h^2\gamma_i^T K\gamma_i - \gamma_i^T M\gamma_{i+1}\right)$$
$$+ \left(-\gamma_{i_0-1}^T M\gamma_{i_0-2} + 2\gamma_{i_0-1}^T M\gamma_{i_0-1} + h^2\gamma_{i_0-1}^T K\gamma_{i_0-1}\right) = \gamma_{i_0-1}^T M\gamma_{i_0}. \qquad [5.46]$$

The left-hand side of [5.46] may be rewritten as:

$$h^2\gamma^T A_{h,r,i_0-1}\gamma = \sum_{i=2}^{i_0-1}(\gamma_i-\gamma_{i-1})^T M(\gamma_i-\gamma_{i-1})+\gamma_{i_0-1}^T M\gamma_{i_0-1}+h^2\sum_{i=1}^{i_0-1}\gamma_i^T K\gamma_i,$$

hence A_{h,r,i_0-1}, which is the restriction of $A_{h,r}$ to the i_0-1 first components, is a non-singular matrix as M and K are positive definite and so the solution of [5.44] $(\gamma_1,\cdots,\gamma_{i_0-1})^T$ is non-zero for $\gamma_i \neq 0$. Also, we can rewrite [5.46] as:

$$\sum_{i=2}^{i_0-1}(\gamma_i-\gamma_{i-1})^T M(\gamma_i-\gamma_{i-1})+h^2\sum_{i=1}^{i_0-1}\gamma_i^T K\gamma_i = \gamma_{i_0-1}^T\left(M\gamma_{i_0} - M\gamma_{i_0-1}\right),$$

and, since $\gamma_{i_0-1} = (I - hP_{i_0})\gamma_{i_0}$, we have that:

$$\gamma_{i_0-1}^T MP_{i_0}\gamma_{i_0} = \gamma_{i_0}^T(I - hP_{i_0})^T MP_{i_0}\gamma_{i_0} > 0 \quad \text{for all } \gamma_{i_0} \neq 0$$

and so:

$$\gamma_{i_0}^T MP_{i_0}\gamma_{i_0} > h\gamma_{i_0}^T P_{i_0}^T MP_{i_0}\gamma_{i_0} = h(P_{i_0}\gamma_{i_0})^T MP_{i_0}\gamma_{i_0} \geq 0 \quad \text{for all } \gamma_{i_0} \neq 0.$$

Similarly, symmetry can be shown by left-multiplying [5.44] by γ'^T and observing that the result does not change when γ and γ' are permuted. □

However, as $u_{h,r}$ is a solution of $(\mathcal{P}_{h,r})$, we have that:

$$Mu_1 + h^2Ku_1 - Mu_2 = h^2 f_1 + h\left(\int_{\mathcal{O}} u_0(y)\varphi_j(y)\mathrm{d}y\right)_{j=1}^{k},$$

$$-Mu_{i-1} + 2Mu_i + h^2Ku_i - Mu_{i+1} = h^2 f_i \quad \forall i \in \{2, ..., N-1\}.$$

Together with the conditions $P_1 = 0$ and $w_{1/2} = -u_{0,r}$, this implies that:

$$Ku_i - f_i = MP_{i+1}\frac{(u_{i+1} - u_i)}{h} + M\frac{(P_{i+1} - P_i)}{h}u_i + M\frac{(w_{i+1/2} - w_{i-1/2})}{h},$$
$$\forall i \in \{1, ..., N-1\},$$

which is a discrete approximation of [1.18].

From [5.45], we have that:

$$u_i = (I - hP_{i+1})u_{i+1} - hw_{i+1/2}, \quad \forall i \in \{0, ..., N-1\}.$$

Thus:

$$\begin{aligned} K(I - hP_{i+1})u_{i+1} - hKw_{i+1/2} - f_i \\ = MP_{i+1}\frac{u_{i+1}}{h} - M\frac{P_i}{h}(I - hP_{i+1})u_{i+1} \\ + MP_iw_{i+1/2} + M\frac{(w_{i+1/2} - w_{i-1/2})}{h}, \end{aligned} \quad [5.47]$$

for $i \in \{1, ..., N-1\}$ with "arbitrary" u_{i+1}. From the terms involving u_{i+1} in [5.47], we obtain the discrete Riccati equation for P:

$$\begin{cases} -M\dfrac{P_{i+1} - P_i}{h}(I - hP_{i+1}) = MP_{i+1}^2 - K(I - hP_{i+1}), \\ \qquad\qquad\qquad\qquad \forall i \in \{1, ..., N-1\} \\ P_1 = 0. \end{cases} \quad [5.48]$$

From this, we deduce that:

$$(M + hMP_i + h^2K)P_{i+1} = MP_i + hK. \quad [5.49]$$

Then, as M, MP_i and K are positive, $(M + hMP_i + h^2K)$ is also positive and we can calculate P_i as follows: $P_1 = 0$ and for $i \in \{1, \cdots, N-1\}$:

$$P_{i+1} = (M + hMP_i + h^2K)^{-1}(MP_i + hK). \quad [5.50]$$

Now, from [5.50], we deduce that:

$$I - hP_{i+1} = (M + hMP_i + h^2K)^{-1}M, \quad [5.51]$$

which we just showed is a non-singular matrix, and so, from [5.48]:

$$\begin{cases} -M\dfrac{P_{i+1} - P_i}{h} = MP_{i+1}^2(I - hP_{i+1})^{-1} - K \ \forall i \in \{1, ..., N-1\}, \\ P_1 = 0. \end{cases}$$

[5.52]

Furthermore, from the independent terms in [5.47], we have that:

$$M\frac{w_{i+1/2} - w_{i-1/2}}{h} = -(MP_i + hK)w_{i+1/2} - f_i.$$

Therefore, from [5.50] and [5.51], we find that:

$$(MP_i + hK) = M(I - hP_{i+1})^{-1}P_{i+1} \qquad [5.53]$$

and, hence:

$$\begin{cases} M\dfrac{w_{i+1/2} - w_{i-1/2}}{h} = -M(I - hP_{i+1})^{-1}P_{i+1}w_{i+1/2} - f_i \\ \qquad\qquad\qquad \forall i \in \{1, ..., N-1\}, \\ w_{1/2} = -u_{0,r}. \end{cases} \qquad [5.54]$$

Additionally, from equation [5.45], we deduce that:

$$\begin{cases} \dfrac{u_{i+1} - u_i}{h} = P_{i+1}u_{i+1} + w_{i+1/2} \quad \forall i \in \{1, ..., N-1\}, \\ u_N = u_{1,r}, \end{cases}$$

where $u_{1,r} \in \mathbb{R}^k$ are the coordinates in V_r of an approximation of u_1.

REMARK 5.5.– We have established the following discretized version of system [1.22–1.24]:

$$\begin{cases} -M\dfrac{P_{i+1} - P_i}{h} = MP_{i+1}^2(I - hP_{i+1})^{-1} - K \;\; \forall i \in \{1, ..., N-1\}, \\ P_1 \equiv 0, \end{cases}$$

$$\begin{cases} M\dfrac{w_{i+1/2} - w_{i-1/2}}{h} = -M(I - hP_{i+1})^{-1}P_{i+1}w_{i+1/2} - f_i \\ \forall i \in \{1, ..., N-1\}, \\ w_{1/2} = -u_{0,r}, \end{cases}$$

$$\begin{cases} \dfrac{u_{i+1} - u_i}{h} = P_{i+1}u_{i+1} + w_{i+1/2} \quad \forall i \in \{1, ..., N-1\}, \\ u_N = u_{1,r}. \end{cases}$$

Using [5.53] and the relation $(I - hP_{i+1})^{-1}P_{i+1} = P_{i+1}(I - hP_{i+1})^{-1}$, the above system may be rewritten equivalently as:

$$\begin{cases} -M\dfrac{P_{i+1} - P_i}{h} = (MP_i + hK)P_{i+1} - K \;\; \forall i \in \{1, ..., N-1\}, \\ P_1 \equiv 0, \end{cases}$$

$$\begin{cases} M\dfrac{w_{i+1/2}-w_{i-1/2}}{h} = -(MP_i + hK)w_{i+1/2} - f_i \ \forall i \in \{1, ..., N-1\}, \\ w_{1/2} = -u_{0,r}, \end{cases}$$

$$\begin{cases} \dfrac{u_{i+1}-u_i}{h} = P_{i+1}u_{i+1} + w_{i+1/2} \quad \forall i \in \{1, ..., N-1\}, \\ u_N = u_{1,r}. \end{cases}$$

Note that the equations for P_i are now linear at each step i of the solution, and once P_i has been calculated, it is not necessary to perform matrix inversion to find the equations for w_i.

We conclude this section with the remark that there exists a wide variety of approaches to the discretized problem. We have shown above that discretizing the problem by finite differences along the x-axis and by finite elements on the section $\mathcal{O}$, and then performing block factorization using a Gaussian method , may be viewed as a discretization of the factorized problem [1.22], [1.23] and [1.24]. We could also perform semi-discretization in y, the factorized problem [1.22], [1.23] and [1.24] by finite elements. This would introduce matrices $MP_h(x)$ such that $(MP_h(x))_{i,j} = (P(x)\varphi_i, \varphi_j)$, and we would naturally recover the fact that the matrices MP_i in proposition 5.1 are symmetric positive definite. We can then choose an appropriate discretization in x.

We could also undertake a global discretization of the cylinder similarly, by means of a finite element method with regular triangulation, working from a fixed triangulation of sections separated by regular intervals. In this case, the elements in the off-diagonal blocks of the matrix $A_{h,r}$ are obtained from the elements of the stiffness matrix associated with the edges connecting pairs of sections. The rest of the calculation is similar to the calculations presented above.

In the case of unstructured finite element methods, the front of the method of invariant embedding based on triangulation is composed of surfaces of arbitrary shape. The continuous factorization calculation must therefore use the equations from Chapter 7.

6

Other Problems

In this chapter, we will continue to consider the spatial domain of problem $(\mathcal{P}_0)$ from section 1.1, which is the cylinder $\Omega =]0, a[\times \mathcal{O}$ in $\mathbb{R}^n$. This will allow us to formulate spatial invariant embedding more simply, but we also wish to show that the factorization method may be extended to other classical types of linear elliptic equations. At the end of the chapter, we will discuss parabolic problems.

6.1. General second-order linear elliptic problems

6.1.1. *Problem statement*

We consider the problem associated with a second-order elliptic operator in the open set Ω. We will continue to use the notation from section 1.1. In particular, x is the coordinate along the axis of the cylinder and is used as an index for this coordinate, and y represents the coordinates within the cylindrical section and so it can also be a multi-index:

$$(\mathcal{P}) \begin{cases} \mathcal{A}u = -\text{div } (A(x,y)\nabla u) + a(x,y)u = f \ \text{ in } \Omega, \\ u|_\Sigma = 0, \quad \dfrac{\partial u}{\partial n_A} = A\nabla u \cdot \vec{n}|_{\Gamma_0} = -a_{xx}\dfrac{\partial u}{\partial x}|_{\Gamma_0} - a_{xy}\nabla_y u|_{\Gamma_0} = u_0, \\ u|_{\Gamma_a} = u_1, \end{cases}$$

where $\vec{n}$ represents the outward unit normal vector.

The matrix $A \in (L^\infty(\Omega))^{n\times n}$ is defined by:

$$A = \begin{pmatrix} a_{xx} & a_{xy} \\ a_{yx} & a_{yy} \end{pmatrix},$$

where $a_{xx} \in \mathbb{R}$, $a_{xy} \in \mathbb{R}^{1\times(n-1)}$, $a_{yx} \in \mathbb{R}^{(n-1)\times 1}$ and $a_{yy} \in \mathbb{R}^{(n-1)\times(n-1)}$, and there exists $\alpha > 0$, such that it satisfies:

$$\vec{\xi}^T A(x,y)\vec{\xi} \geq \alpha \left\|\vec{\xi}\right\|^2, \quad \forall \vec{\xi} \in \mathbb{R}^n.$$

We assume that $a \in L^\infty(\Omega) \geq a_0 > 0$. For this problem, we choose $f \in L^2(\Omega)$, $u_0 \in H^{1/2}_{0,0}(\mathcal{O})'$ and $u_1 \in H^{1/2}_{0,0}(\mathcal{O})$. Using the Green's formula, the variational formulation of the problem may be written as follows:

$$\int_\Omega (A\nabla u\cdot\nabla v + auv)\,\mathrm{d}x\,\mathrm{d}y = (f,v) + < u_0, v|_{\Gamma_0} >_{H^{1/2}_{0,0}(\mathcal{O})'\times H^{1/2}_{0,0}(\mathcal{O})},$$
$$\forall v \in X_0, \quad [6.1]$$

with $u|_{\Gamma_a} = u_1$. Then, problem $(\mathcal{P})$ is well-posed and it admits a unique solution in $H^1(\Omega)$.

6.1.2. *Factorization by invariant embedding*

We use the method of spatial invariant embedding introduced in Chapter 1: we embed problem $(\mathcal{P})$ into a family of similar problems $(\mathcal{P}_{s,h})$ defined on $\Omega_s =]0,s[\times\mathcal{O}$ for $s \in]0,a]$. For each problem, we impose a Dirichlet boundary condition $u_s|_{\Gamma_s} = h$, where $\Gamma_s = \{s\}\times\mathcal{O}$:

$$(\mathcal{P}_{s,h})\begin{cases} \mathcal{A}u_s = f & \text{in } \Omega_s, \\ u_s|_\Sigma = 0, & \dfrac{\partial u_s}{\partial n_{\mathcal{A}}}|_{\Gamma_0} = u_0, \quad u_s|_{\Gamma_s} = h. \end{cases} \quad [6.2]$$

Problem $(\mathcal{P}_{s,h})$ has a unique solution in $X_s = L^2(0,s;H^1_0(\mathcal{O})) \cap H^1(0,s;L^2(\mathcal{O}))$. It is clear that $(\mathcal{P})$ corresponds exactly to $(\mathcal{P}_{s,h})$ for $s = a, h = u_1$.

For $s \in]0,a]$, we define $Y_s = \{v \in H^1(\Omega_s) : \dfrac{\partial^2 v}{\partial x^2} \in L^2(0,s;H^{-1}(\mathcal{O}))$ and $v|_\Sigma = 0\}$. The calculation is now similar to that given in section 1.2.

DEFINITION 6.1.– For all $s \in]0, a]$, we define the Dirichlet-to-Neumann (DtN) map $P(s)$ by $P(s)h = \frac{\partial \gamma_s}{\partial n_{\mathcal{A}}}|_{\Gamma_s} = A\nabla\gamma_s \cdot \vec{n}|_{\Gamma_s} = a_{xx}\frac{\partial \gamma_s}{\partial x}|_{\Gamma_s} + a_{xy}\nabla_y\gamma_s|_{\Gamma_s}$, where $h \in H^{1/2}_{0,0}(\mathcal{O})$ and $\gamma_s \in Y_s$ is the solution of:

$$\begin{cases} \mathcal{A}\gamma_s = 0 & \text{in } \Omega_s, \\ \gamma_s|_\Sigma = 0, & \frac{\partial \gamma_s}{\partial n_{\mathcal{A}}}|_{\Gamma_0} = 0, \quad \gamma_s|_{\Gamma_s} = h. \end{cases} \qquad [6.3]$$

We set $P(0)h = 0$ in reference to the condition on Γ_0 and define w by $w(s) = \frac{\partial \beta_s}{\partial n_{\mathcal{A}}}|_{\Gamma_s}$, where $\beta_s \in Y_s$ is the solution of:

$$\begin{cases} \mathcal{A}\beta_s = f & \text{in } \Omega_s, \\ \beta_s|_\Sigma = 0, & \frac{\partial \beta_s}{\partial n_{\mathcal{A}}}|_{\Gamma_0} = u_0, \quad \beta_s|_{\Gamma_s} = 0. \end{cases} \qquad [6.4]$$

We also set $w(0) = -u_0$, again referring to the condition on Γ_0.

The following proposition makes precise the meaning of these definitions.

PROPOSITION 6.1.– The linear operator $P(s) : H^{1/2}_{0,0}(\mathcal{O}) \mapsto H^{1/2}_{0,0}(\mathcal{O})'$ is continuous and positive for all $s \in]0, a]$ and the associated bilinear form on $H^{1/2}_{0,0}(\mathcal{O})$, $< P(s)h, \overline{h} >_{H^{1/2}_{0,0}(\mathcal{O})' \times H^{1/2}_{0,0}(\mathcal{O})}$, is coercive for all $s \in]0, a]$. The function $w(s)$ is defined in $H^{1/2}_{0,0}(\mathcal{O})'$ for all s.

PROOF.– Let $h, \overline{h} \in H^{1/2}_{0,0}(\mathcal{O})$ and $\gamma_s, \overline{\gamma}_s$ be the corresponding solutions in Ω_s. By the Green's formula, we have that:

$$0 = \int_{\Omega_s} \left(A\nabla\gamma_s \cdot \nabla\overline{\gamma}_s + a(x,y)\gamma_s\overline{\gamma}_s\right) \mathrm{d}x\,\mathrm{d}y - < P(s)h, \overline{h} >_{H^{1/2}_{0,0}(\mathcal{O})' \times H^{1/2}_{0,0}(\mathcal{O})} .$$

We deduce that $P(s)$ is continuous from $H^{1/2}_{0,0}(\mathcal{O})$ to $H^{1/2}_{0,0}(\mathcal{O})'$. If we set $h = \overline{h}$, by the Poincaré inequality and positivity of the matrix A, we can show the following inequality in a similar manner to proposition 1.2:

$$< P(s)h, \overline{h} >_{H^{1/2}_{0,0}(\mathcal{O})' \times H^{1/2}_{0,0}(\mathcal{O})} \geq C_s \|\nabla\gamma_s\|^2_{L^2(\Omega_s)} \geq C'_s \|h\|^2_{H^{1/2}_{0,0}(\mathcal{O})},$$

which proves coercivity. By extending the variational formulation [6.1] to non-zero test functions on Γ_s, we have the following for β_s:

$$\int_{\Omega_s} f\overline{\beta}_s \,\mathrm{d}x\,\mathrm{d}y - \int_{\Omega_s} \left(A\nabla\beta_s \cdot \nabla\overline{\beta}_s + a(x,y)\beta_s\overline{\beta}_s\right) \mathrm{d}x\,\mathrm{d}y$$
$$+ < u_0, \overline{\beta}_s|_{\Gamma_0} >_{H^{1/2}_{0,0}(\mathcal{O})' \times H^{1/2}_{0,0}(\mathcal{O})} =< w(s), \overline{\beta}_s|_{\Gamma_s} >_{H^{1/2}_{0,0}(\mathcal{O})' \times H^{1/2}_{0,0}(\mathcal{O})} \quad [6.5]$$

for all $\overline{\beta}_s$ in X_s, which defines $w(s) \in H^{1/2}_{0,0}(\mathcal{O})'$ as required. □

Therefore, by linearity of $(\mathcal{P}_{s,h})$ and using the appropriate matrix product for each of the terms:

$$\frac{\partial u_s}{\partial n_A}|_{\Gamma_s} = A\nabla u_s \vec{n}|_{\Gamma_s} = a_{xx}\frac{\partial u_s}{\partial x}|_{\Gamma_s} + a_{xy}\nabla_y u_s|_{\Gamma_s} = P(s)h + w(s), \quad [6.6]$$

where $u_s = \beta_s + \gamma_s$, and the index s reminds us that the problem is defined on Ω_s.

For $x \leq s$, we now consider the restriction of $u_s(h)$ to Ω_x. This is a solution of problem $(\mathcal{P}_{x,u_s(x;h)})$. By the above argument, we therefore have the relation:

$$a_{xx}(x,\cdot)\frac{\partial u_s}{\partial x}(x;h) + a_{xy}(x,\cdot)\nabla_y u_s(x;h) = P(x)u_s(x;h) + w(x). \quad [6.7]$$

Returning to the analogy of evolution problems, we refer to the following maps as trajectories:

$$x \longmapsto u_s(x;h) = u_s(x,\cdot;h) \in H^{1/2}_{0,0}(\mathcal{O}).$$

Thus, [6.7] is true for the family of trajectories depending on h. Each trajectory is not only continuous on $[0,a]$ with values in $H^{1/2}_{0,0}(\mathcal{O})$, but also defined in $L^2(0,a;H^1(\mathcal{O}))$. We can therefore, at least in a formal sense, take the derivative of [6.7] w.r.t. x, and using the equation satisfied by u_s:

$$\frac{\partial}{\partial x}\left(a_{xx}\frac{\partial u_s}{\partial x} + a_{xy}\nabla_y u_s\right) =$$
$$= -\mathrm{div}_y\left(a_{yx}\frac{\partial u_s}{\partial x} + a_{yy}\nabla_y u_s\right) + a u_s - f = \frac{\mathrm{d}P}{\mathrm{d}x}u_s + P\frac{\partial u_s}{\partial x} + \frac{\mathrm{d}w}{\mathrm{d}x}. \quad [6.8]$$

By substituting $\frac{\partial u_s}{\partial x}$ using [6.7], we get:

$$-\operatorname{div}_y\left(a_{yx}\left(a_{xx}^{-1}Pu_s + a_{xx}^{-1}w - a_{xx}^{-1}a_{xy}\nabla_y u_s\right) + a_{yy}\nabla_y u_s\right) + au_s - f$$
$$= \frac{\mathrm{d}P}{\mathrm{d}x}u_s + P\left(a_{xx}^{-1}Pu_s + a_{xx}^{-1}w - a_{xx}^{-1}a_{xy}\nabla_y u_s\right) + \frac{\mathrm{d}w}{\mathrm{d}x}. \quad [6.9]$$

Then:

$$\Big(\frac{\mathrm{d}P}{\mathrm{d}x} + Pa_{xx}^{-1}P - Pa_{xx}^{-1}a_{xy}\nabla_y + \mathrm{div}_y a_{yx}a_{xx}^{-1}P - \mathrm{div}_y a_{yx}a_{xx}^{-1}a_{xy}\nabla_y$$
$$+ \mathrm{div}_y a_{yy}\nabla_y - a\Big)(x)u_s(x)$$
$$+ \left(\frac{\mathrm{d}w}{\mathrm{d}x} + Pa_{xx}^{-1}w + \mathrm{div}_y a_{yx}a_{xx}^{-1}w - V_x a_{xx}^{-1}w + f\right)(x) = 0. \quad [6.10]$$

The above equality is satisfied for all $x \in [0, s]$. By setting $x = s$, we find that:

$$\Big(\frac{\mathrm{d}P}{\mathrm{d}x} + Pa_{xx}^{-1}P - Pa_{xx}^{-1}a_{xy}\nabla_y + \mathrm{div}_y a_{yx}a_{xx}^{-1}P - \mathrm{div}_y a_{yx}a_{xx}^{-1}a_{xy}\nabla_y +$$
$$\mathrm{div}_y a_{yy}\nabla_y - a\Big)(s)h$$
$$+ \left(\frac{\mathrm{d}w}{\mathrm{d}x} + Pa_{xx}^{-1}w + \mathrm{div}_y a_{yx}a_{xx}^{-1}w - V_x a_{xx}^{-1}w + f\right)(s) = 0. \quad [6.11]$$

This calculation holds for any trajectory satisfying the equation $\mathcal{A}u = f$ and the boundary condition on Γ_0, $A\nabla u \cdot \vec{n}|_{\Gamma_0} = u_0$. The function u is fully determined if h and s are given. Since h and s are arbitrary, we deduce that:

$$\frac{\mathrm{d}P}{\mathrm{d}x} + Pa_{xx}^{-1}P - Pa_{xx}^{-1}a_{xy}\nabla_y + \mathrm{div}_y a_{yx}a_{xx}^{-1}P$$
$$= \mathrm{div}_y a_{yx}a_{xx}^{-1}a_{xy}\nabla_y - \mathrm{div}_y a_{yy}\nabla_y + a, \quad \text{for } x \in]0, a], \quad [6.12]$$

and for the residual term:

$$\frac{\mathrm{d}w}{\mathrm{d}x} + Pa_{xx}^{-1}w + \mathrm{div}_y a_{yx}a_{xx}^{-1}w = -f, \quad \text{for } x \in]0, a]. \quad [6.13]$$

We use the same reasoning as in the case of the Poisson equation, extending by continuity to $x = 0$ and arguing that the trajectory is arbitrary. Using the fact that $u(0)$ is arbitrary and applying relation [6.7] at $x = 0$, we obtain the initial conditions for P and w:

$$P(0) = 0, \quad w(0) = -u_0. \tag{6.14}$$

We wish to emphasize the fact that P is an operator dependent on x that acts on functions in the variable y defined on the section $\mathcal{O}$. It satisfies a Riccati equation. The right-hand side is an unbounded operator on $L^2(\mathcal{O})$. Regrouping equations [6.12], [6.13], the initial conditions for P and w, equation [6.7], and the condition on Γ_a for u, we obtain the system:

$$\begin{cases} \dfrac{\mathrm{d}P}{\mathrm{d}x} + Pa_{xx}^{-1}P - Pa_{xx}^{-1}a_{xy}\nabla_y + \mathrm{div}_y a_{yx}a_{xx}^{-1}P \\ \qquad = \mathrm{div}_y a_{yx}a_{xx}^{-1}a_{xy}\nabla_y - \mathrm{div}_y a_{yy}\nabla_y + a, \qquad P(0) = 0, \\ \dfrac{\mathrm{d}w}{\mathrm{d}x} + Pa_{xx}^{-1}w + \mathrm{div}_y a_{yx}a_{xx}^{-1}w = -f, \qquad w(0) = -u_0, \\ -a_{xx}\dfrac{\mathrm{d}u}{\mathrm{d}x} - a_{xy}\nabla_y u + Pu = -w(x), \qquad u(a) = u_1. \end{cases} \tag{6.15}$$

REMARK 6.1.– If the matrix A is symmetric, we can write the first equation of system [6.15] as:

$$\frac{\mathrm{d}P}{\mathrm{d}x} + Pa_{xx}^{-1}P + PB^{\mathrm{T}} + BP = \mathrm{div}_y a_{yx}a_{xx}^{-1}a_{xy}\nabla_y - \mathrm{div}_y a_{yy}\nabla_y + a,$$

with $B = \mathrm{div}_y a_{yx}a_{xx}^{-1}$, from which we deduce self-adjointness of P. Now, we observe that terms linear in P have appeared in the Riccati equation. In the case of the Laplacian, the right-hand side is the restriction of this operator to the section $\mathcal{O}$, that is, considering functions constant in x. In this case, the term appearing in the right-hand side is obtained from the operator $\mathcal{A}$ applied to functions constant in the direction of the conormal $n_{\mathcal{A}}$.

REMARK 6.2.– System [6.15] is decoupled, as the first two equations can be integrated w.r.t. x from 0 to a to obtain P and w ("upward phase"), then u is obtained by backward integration of the third equation ("downward phase"). Formally, once P is known, $(\mathcal{P})$ factorizes into:

$$-\left(\frac{\mathrm{d}}{\mathrm{d}x} + Pa_{xx}^{-1} + \mathrm{div}_y a_{yx}a_{xx}^{-1}\right)\left(a_{xx}\frac{\mathrm{d}}{\mathrm{d}x} + a_{xy}\nabla_y - P\right)u = f. \tag{6.16}$$

REMARK 6.3.– This factorization may be generalized to other strongly elliptic operators. Consider, for example, problem $(\mathcal{P})$ with an added first-order convective term $B \cdot \nabla u$, such that $B \in (L^\infty(\Omega))^{1\times n}$ and $\mathrm{div} B = 0$. In this case, we know that the corresponding boundary value problem is coercive. We then have the factorization:

$$\left\{\begin{array}{ll} \dfrac{\mathrm{d}P}{\mathrm{d}x} + Pa_{xx}^{-1}P - Pa_{xx}^{-1}a_{xy}\nabla_y + \mathrm{div}_y a_{yx}a_{xx}^{-1}P - B_x a_{xx}^{-1}P & \\ = \mathrm{div}_y a_{yx}a_{xx}^{-1}a_{xy}\nabla_y - \mathrm{div}_y a_{yy}\nabla_y - B_x a_{xx}^{-1}a_{xy}\nabla_y + B_y\nabla_y + a, & P(0) = 0, \\ \dfrac{\mathrm{d}w}{\mathrm{d}x} + Pa_{xx}^{-1}w + \mathrm{div}_y a_{yx}a_{xx}^{-1}w - B_x a_{xx}^{-1}w = -f, & w(0) = -u_0, \\ -a_{xx}\dfrac{\mathrm{d}u}{\mathrm{d}x} - a_{xy}\nabla_y u + Pu = -w(x), & u(a) = u_1. \end{array}\right.$$

6.2. Systems of coupled boundary value problems

We consider the following problem, which provides a model for systems of elliptic boundary value problems:

$$-\alpha_1\Delta u_1 + \alpha_{1,2}(x)u_2 = f_1 \quad \text{in } \Omega, \qquad [6.17]$$

$$-\alpha_2\Delta u_2 + \alpha_{2,1}(x)u_1 = f_2 \quad \text{in } \Omega, \qquad [6.18]$$

$$u_1|_\Sigma = 0, \quad u_2|_\Sigma = 0, \qquad [6.19]$$

$$-\alpha_1\frac{\partial u_1}{\partial x}|_{\Gamma_0} = u_{1,0}, \quad -\alpha_2\frac{\partial u_2}{\partial x}|_{\Gamma_0} = u_{2,0}, \qquad [6.20]$$

$$u_1|_{\Gamma_a} = u_{1,1} \quad u_2|_{\Gamma_a} = u_{2,1}, \qquad [6.21]$$

with $\alpha_i > 0, \alpha_i \in L^\infty(\Omega), u_{i,0} \in (H_{0,0}^{1/2}(\mathcal{O}))', u_{i,1} \in H_{0,0}^{1/2}(\mathcal{O})$ for $i = 1, 2$.

The notation for this domain is the same as in section 1.1.

The main purpose of this section is to exhibit a variety of different factorization methods. We will demonstrate two factorization methods according to whether the unknown functions on each section are handled globally (global approach) or invariant embedding is performed on each of the unknown functions successively (sequential approach). By referring to the discretized problem (see Chapter 5), we can draw a connection between these methods and the different approaches to indexing unknowns: the unknowns u_1 and u_2 can be either indexed successively at each node, or we can list all unknowns u_1 followed by all unknowns u_2.

Here, we again use definition 1.1, extending it to apply to systems: let X and X_0 be the Hilbert spaces defined by:

$$X = (L^2(0, a; (H_0^1(\mathcal{O}))^2 \cap (H^1(0, a; (L^2(\mathcal{O})))^2,$$
$$X_0 = X \cap \{(u_1, u_2) \in X : \; u_1|_{\Gamma_a} = 0 \; u_2|_{\Gamma_a} = 0\}.$$

The variational formulation of [6.17–6.21] can be written as follows for all $(\overline{u_1}, \overline{u_2}) \in X_0$:

$$\alpha_1(\nabla u_1, \nabla\overline{u_1}) + \alpha_2(\nabla u_2, \nabla\overline{u_2}) + (\alpha_{1,2}u_1, \overline{u_2}) + (\alpha_{2,1}u_2, \overline{u_1}) = (f_1, \overline{u_1})$$
$$+ (f_2, \overline{u_2}) + < u_{1,0}, \overline{u_1}|_{\Gamma_0} >_{H_{0,0}^{1/2}(\mathcal{O})' \times H_{0,0}^{1/2}(\mathcal{O})}$$
$$+ < u_{2,0}, \overline{u_2}|_{\Gamma_0} >_{H_{0,0}^{1/2}(\mathcal{O})' \times H_{0,0}^{1/2}(\mathcal{O})} . \qquad [6.22]$$

Let C be the constant in the Poincaré inequality of X_0:

$$\|u\|_{L^2(\Omega)} \leq C \, \|\nabla u\|_{L^2(\Omega)} .$$

The bilinear form in [6.22] is coercive, and so the existence and uniqueness of the solution of [6.17–6.21] follows from the Lax–Milgram theorem whenever:

$$\frac{C}{2} \sup_x(|\alpha_{1,2}| + |\alpha_{2,1}|) < \inf\{\alpha_1, \alpha_2\}. \qquad [6.23]$$

6.2.1. *Global approach*

We apply the method of invariant embedding described in section 1.2 to the couple of functions (u_1, u_2). On the variable section Γ_s, we have:

$$\begin{pmatrix} \alpha_1 \dfrac{\partial u_1}{\partial x} \\ \alpha_2 \dfrac{\partial u_2}{\partial x} \end{pmatrix} = \begin{pmatrix} P_{11} & P_{12} \\ P_{21} & P_{22} \end{pmatrix} \begin{pmatrix} u_1 \\ u_2 \end{pmatrix} + \begin{pmatrix} w_1 \\ w_2 \end{pmatrix} . \qquad [6.24]$$

By setting:

$$u = \begin{pmatrix} u_1 \\ u_2 \end{pmatrix}, \quad w = \begin{pmatrix} w_1 \\ w_2 \end{pmatrix}, \quad P = \begin{pmatrix} P_{11} & P_{12} \\ P_{21} & P_{22} \end{pmatrix},$$
$$J_\alpha = \begin{pmatrix} 0 & \alpha_{1,2} I \\ \alpha_{2,1} I & 0 \end{pmatrix}, \quad A = \begin{pmatrix} \alpha_1 & 0 \\ 0 & \alpha_2 \end{pmatrix}, \qquad [6.25]$$

we obtain a factorization that may be written as follows:

$$\frac{\mathrm{d}P}{\mathrm{d}x} + PA^{-1}P = -\begin{pmatrix} \alpha_1 \Delta_y & 0 \\ 0 & \alpha_2 \Delta_y \end{pmatrix} + J_\alpha, \qquad P(0) = 0, \qquad [6.26]$$

$$\frac{\mathrm{d}w}{\mathrm{d}x} + PA^{-1}w = -f, \qquad w(0) = -u_0, \qquad [6.27]$$

$$-A\frac{\mathrm{d}u}{\mathrm{d}x} + Pu = -w, \qquad u(a) = u_1. \qquad [6.28]$$

6.2.2. *Sequential approach*

Consider now a second method in which we will successively apply invariant embedding to u_1 and then to u_2. We can think of this approach as duplicating the domain Ω into two domains Ω_1 and Ω_2 on which u_1 and u_2 are, respectively, defined. The interaction between these two functions is therefore non-local. Invariant embedding is achieved by successively sweeping across both domains. We will however continue to write x as the longitudinal coordinate in each of the cylinders ($0 \leq x \leq a$). Thus, first consider the equation in u_1 where u_2 is assumed to be known, and apply the factorization presented in section 3.3 with invariant embedding for $0 \leq x \leq a$, which explicitly yields the dependency of u_1 on u_2. The normal derivative of u_1 at x is an affine function of the (arbitrary) data $u_1(x)$ and $u_2(x'), 0 \leq x' \leq x$:

$$-\alpha_1 \frac{\mathrm{d}u_1}{\mathrm{d}x} + P_1 u_1 - \int_0^x R_2(x', x) u_2(x')\, \mathrm{d}x' = -w_1, \quad u_1(a) = u_{1,1}. \qquad [6.29]$$

By differentiating [6.29] w.r.t. x, we obtain:

$$\left(\frac{\mathrm{d}P_1}{\mathrm{d}x} + P_1\frac{1}{\alpha_1}P_1 + \alpha_1\Delta_y\right)u_1 \\ - \int_0^x \left(\frac{\partial R_2}{\partial x}(x', x) + P_1(x)\frac{1}{\alpha_1}R_2(x', x)\right)u_2(x')\,\mathrm{d}x' \\ - \alpha_{1,2}(x)u_2(x) - R_2(x, x)u_2(x) = -\frac{\mathrm{d}w_1}{\mathrm{d}x} - P_1\frac{1}{\alpha_1}w_1 - f_1. \qquad [6.30]$$

Noting that $u_1(x)$, $u_2(x)$ and also $u_2(x'), x \leq x' \leq a$ are arbitrary, we deduce that the operator P_1 satisfies the Riccati equation:

$$\frac{\mathrm{d}P_1}{\mathrm{d}x} + P_1\frac{1}{\alpha_1}P_1 = -\alpha_1\Delta_y, \quad P_1(0) = 0, \qquad [6.31]$$

and for the residual:

$$\frac{\mathrm{d}w_1}{\mathrm{d}x} + P_1\frac{1}{\alpha_1}w_1 = -f_1, \quad w_1(0) = -u_0^1, \qquad [6.32]$$

and the operator $R_2(x', x)$ satisfies the analogous counterpart of [3.43]:

$$\frac{\partial R_2}{\partial x}(x', x) + P_1(x)\frac{1}{\alpha_1}R_2(x', x) = 0, \quad x' \leq x \leq a, \\ R_2(x', x') = -\alpha_{1,2}(x')I. \qquad [6.33]$$

Here, the initial condition on R_2 expresses the coupling with the equation for u_2.

Now, we perform invariant embedding once again by considering u_1 determined by equation [6.17] on the whole cylinder Ω, and then again on the sub-cylinder Ω_s for u_2 as defined by equation [6.18]. The unknowns are therefore given by the couple u_1 in Ω and $u_{2,s}$ in Ω_s. Thus, we consider the invariant embedding for $0 \leq s \leq a$: to calculate $u_{2,s}$ in the sub-cylinder, we fix u_2 on Γ_s, but we also need to know $u_2(x)$ for $s \leq x \leq a$ in order to solve equation [6.17] in the cylinder Ω. We therefore have:

$$-\alpha_2\frac{\mathrm{d}u_2}{\mathrm{d}x} + P_2u_2 - \int_x^a R_1(x', x)u_2(x')\,\mathrm{d}x' = -w_2, \quad u_2(a) = u_{2,1}. \qquad [6.34]$$

As $\dfrac{du_2}{dx}$ is fixed at $x = 0$, it is independent of $u_2(0)$ and $u_2(x)$. Hence, we have the initial conditions:

$$P_2(0) = 0, \quad R_1(x', 0) = 0. \qquad [6.35]$$

We differentiate this expression w.r.t. x and apply relation [6.34] together with equation [6.18]:

$$\begin{aligned}&\left(\frac{dP_2}{dx} + P_2\frac{1}{\alpha_2}P_2 + \alpha_2\Delta_y + R_1(x,x)\right)u_2\\&-\int_x^a\left(\frac{\partial R_1(x',x)}{\partial x} + P_2\frac{1}{\alpha_2}R_1(x',x)\right)u_2(x')\,dx'\\&= -\frac{dw_2}{dx} - P_2\frac{1}{\alpha_2}w_2 - f_2 + \alpha_{2,1}u_1.\end{aligned} \qquad [6.36]$$

Here, unlike in the previous calculation, u_1 is not arbitrary. We now establish an equation for w_2:

$$\begin{aligned}&\frac{dw_2}{dx} + P_2\frac{1}{\alpha_2}w_2 + \int_0^x R_1{}'(x,x')\frac{1}{\alpha_2}w_2(x')\,dx'\\&+\int_x^a R_2{}'(x,x')\frac{1}{\alpha_1}w_1(x')\,dx' = -f_2,\\&w_2(0) = -u_{2,0}.\end{aligned} \qquad [6.37]$$

We will formally show that R_1, $R_1{}'$ and $R_2{}'$ may be determined, such that if u_2 is a solution of [6.34] then it is also a solution of [6.18]. Note here that the argument that we have applied several times regarding the "arbitrary" nature of $u(x)$ is based on the possible variations of this value and so cannot be used at $x = a$. For the purpose of this argument, we therefore assume that $u_1(a) = u_2(a) = 0$.

We can make the two integral terms in [6.37] explicit by using [6.29] and [6.34]:

$$\int_0^x {R_1}'(x,x')\frac{1}{\alpha_2}w_2(x')\,\mathrm{d}x' = \int_0^x {R_1}'(x,x')\left(\frac{\mathrm{d}u_2}{\mathrm{d}x'}(x') - \frac{1}{\alpha_2}P_2(x')u_2(x')\right.$$

$$\left. + \int_{x'}^a \frac{1}{\alpha_2}R_1(x'',x')u_2(x'')\,\mathrm{d}x''\right)\mathrm{d}x'$$

$$= -\int_0^x \frac{\partial {R_1}'}{\partial x'}(x,x')u_2(x')\,\mathrm{d}x' + {R_1}'(x,x)u_2(x)$$

$$-\int_0^x \frac{1}{\alpha_2}{R_1}'(x,x')P_2(x')u_2(x')\,\mathrm{d}x'$$

$$+\int_0^x\int_0^{x''}\frac{1}{\alpha_2}{R_1}'(x,x')R_1(x'',x')\,\mathrm{d}x'u_2(x'')\,\mathrm{d}x''. \qquad [6.38]$$

$$\int_x^a {R_2}'(x,x')\frac{1}{\alpha_1}w_1(x')\,\mathrm{d}x' = \int_x^a {R_2}'(x,x')\left(\frac{\mathrm{d}u_1}{\mathrm{d}x'}(x') - \frac{1}{\alpha_1}P_1(x')u_1(x')\right.$$

$$\left. + \int_0^{x'} R_2(x'',x')u_2(x'')\,\mathrm{d}x''\right)\mathrm{d}x'$$

$$= -\int_x^a \frac{\partial {R_2}'}{\partial x'}(x,x')u_1(x')\,\mathrm{d}x' + {R_2}'(x,x)u_1(x)$$

$$-\int_x^a \frac{1}{\alpha_1}{R_2}'(x,x')P_1(x')u_1(x')\,\mathrm{d}x'$$

$$+\int_0^a\int_{x\wedge x''}^a\frac{1}{\alpha_1}{R_2}'(x,x')R_2(x'',x')\,\mathrm{d}x'u_2(x'')\,\mathrm{d}x''. \qquad [6.39]$$

If we substitute these expressions into [6.37], we can calculate $\frac{\mathrm{d}w_2}{\mathrm{d}x} + P_2\frac{1}{\alpha_2}w_2$ as a function of $u_1(x'), x \le x' \le a$, $u_2(x)$ and $u_2(x'), 0 \le x' \le a$. We then substitute this value into [6.36]. We thus obtain an expression written in terms of x in the second domain with the following terms: $u_1(x), u_2(x), u_1(x'), x \le x' \le a, u_2(x'), 0 \le x' \le a$. We will choose $R_1, {R_1}', {R_2}'$ and P_2 (bearing in mind that P_1 is already determined by [6.30] and R_2 by [6.33]) in order to identify this relation with equation [6.18] at x.

Consider first the terms in $u_1(x')$ for $x' \geq x$. From these, we deduce that:

$$\frac{\partial R_2{}'}{\partial x'}(x,x') + \frac{1}{\alpha_1} R_2{}'(x,x')P_1(x') = 0, \qquad [6.40]$$
$$R_2{}'(x,x) = -\alpha_{2,1}(x)I.$$

Now, consider the terms in $u_2(x')$ for $x' \geq x$, which give:

$$\frac{\partial R_1{}'}{\partial x'}(x,x') + \frac{1}{\alpha_2} R_1{}'(x,x')P_1(x') - \int_x^a \frac{1}{\alpha_1} R_2{}'(x,x'')R_2(x',x'')\,\mathrm{d}x''$$
$$- \int_0^{x'} \frac{1}{\alpha_2} R_1{}'(x,x'')R^1(x',x'')\,\mathrm{d}x'' = 0, \quad [6.41]$$

$$R_1{}'(x,0) = 0. \qquad [6.42]$$

For the terms in $u_2(x')$ for $x' \leq x$, we have:

$$\frac{\partial R_1}{\partial x}(x',x) + P_2(x)\frac{1}{\alpha_2} R_1(x',x) - \int_{x'}^a \frac{1}{\alpha_1} R_2{}'(x,x'')R_2(x',x'')\,\mathrm{d}x'' = 0,$$
[6.43]

$$R_1(x',0) = 0. \qquad [6.44]$$

Finally, the term in $u_2(x)$ yields the Riccati equation for P_2:

$$\frac{\mathrm{d}P_2}{\mathrm{d}x} + P_2 \frac{1}{\alpha_2} P_2 = -\alpha_2 \Delta_y, \quad P_2(0) = 0. \qquad [6.45]$$

Using these equations, we can check that [6.29] is indeed recovered.

These equations may be solved in the following way:

– solve the two Riccati equations [6.31], [6.45] for P_1 and P_2 from $x = 0$ to $x = a$;

– solve equation [6.33] for R_2 at x ranging from x' to a for each x' from 0 to a; do the same for [6.40], which is an equation in x' ranging from x to a;

– next, in equation [6.43] for R_1 at x' for $x \leq x' \leq a$, R_2 and $R_2{}'$ are known;

– finally, since R_1 is known, [6.41] is an integro-differential equation in x', which may be shown to be well-posed;

– once the factorization of the system has been established by using the operators $P_1, P_2, R_1, R_1{}', R_2, R_2{}'$, for each set of data $f_1, f_2, u_{1,0}, u_{2,0}, u_{1,1}, u_{2,1}$ the solution of [6.17], [6.18] is obtained by successively solving the uncoupled differential or integro-differential equations [6.32], [6.37] and then [6.29] and [6.34] in the backward direction.

REMARK 6.4.– Fill-in is a well-known phenomenon in numerical linear algebra: for sparse matrices, we define the profile as the set of indices of the non-zero terms furthest from the diagonal. In the sequential approach considered here, the profile is determined by the coupling coefficients $\alpha_{1,2}$ and $\alpha_{2,1}$. We know that, after LU factorization, the upper and lower triangular factors inherit this profile, but that the coefficients in the interior of the profile are generally non-zero: this is fill-in. In the case that we examined, the operators $R_1, R_1{}', R_2, R_2{}'$ and the equations that they satisfy provide a description of the fill-in phenomenon. Of course, this approach consequently requires more calculations than the first method presented above, which performs the calculation by the intermediary of the Riccati equation [6.26]. However, this approach allows one of the unknowns to be eliminated, as shown in the following calculation.

6.2.2.1. *Elimination of an unknown*

We will show that this approach allows the unknown u_1 to be eliminated, which leads to an equation in u_2 only. We can obtain u_1 by the following calculation, which is analogous to that performed for formula [3.50]:

$$u_1(x) = S_1(a, x)u_{1,1} - \int_x^a S_1(x', x)\left(-w_1(x') + \int_0^{x'} R_2(t, x')u_2(t)\,\mathrm{d}t\right)\mathrm{d}x', \quad [6.46]$$

where S_1 is the evolution operator in the direction of decreasing x associated with [6.29]:

$$-\alpha_1 \frac{\partial S_1}{\partial x}(x', x) + P_1(x)S_1(x', x) = 0,$$

$$0 \leq x' \leq x, \quad \alpha_1 S_1(x, x) = -I. \quad [6.47]$$

Now u_2 satisfies the following well-posed problem that no longer involves u_1:

$$-\alpha_2 \Delta u_2 + \alpha_{2,1}(x)\,(S_1(a,x)u_{1,1}$$

$$-\int_x^a S_1(x',x)\left(-w_1(x') + \int_0^{x'} R_2(t,x')u_2(t)\,\mathrm{d}t\right)\mathrm{d}x'\right) = f_2 \quad \text{in } \Omega,$$

$$-\alpha_2 \frac{\partial u_2}{\partial x}|_{\Gamma_0} = u_{2,0}, \quad u_2|_\Sigma = 0, \quad u_2|_{\Gamma_a} = u_{2,1},$$

where w_1 is obtained by solving [6.32].

REMARK 6.5.– In the above calculation, we only require the equation to be linear in u_2: the same calculation would hold for a nonlinear equation in u_2.

REMARK 6.6.– Calculating the operators P_2, R_1, R_2', R_1' becomes necessary only if we wish to continue the calculation by eliminating u_2 on a region of Ω.

6.3. Linear elasticity system

We consider a classical problem from continuum mechanics: the study of small deformations of an object whose material parameters are governed by a linear elasticity law. $\mathbb{R}^3$ is therefore the natural setting of this problem, and we adopt the classical formulations used in mechanics in this section, writing the coordinates as (x_1, x_2, x_2).

6.3.1. *Problem statement and transformation*

We consider the cylinder Ω in $\mathbb{R}^3$ with axis parallel to the x_1 axis. Let $\mathcal{O}$ denote the section of this cylinder lying in the plane (x_2, x_3). We assume that this section is a regular bounded open set. The cylinder is defined by $\Omega =]0, a[\times \mathcal{O}$ and its faces are given by $\Gamma_0 = \{0\} \times \mathcal{O}$, $\Gamma_a = \{a\} \times \mathcal{O}$. The lateral surface is $\Sigma =]0, a[\times \partial\mathcal{O}$. Consider the classical system of linear elasticity for a displacement $u = (u_1, u_2, u_3)^T$. We write σ as the stress tensor, and σ_i its i-th column:

$$\sigma_i = (\sigma_{1i}, \sigma_{2i}, \sigma_{3i})^T.$$

Writing $\mathcal{E}$ for the elasticity operator, we consider the problem:

$$(\mathcal{PE})\begin{cases} -\operatorname{div}\sigma = -\mu\Delta u - (\lambda+\mu)\operatorname{grad}\operatorname{div} u = -\mathcal{E}u = f \quad \text{in } \Omega, \\ u|_\Sigma = 0, \\ u|_{\Gamma_a} = u_a, \\ -\sigma_1|_{\Gamma_0} = g. \end{cases}$$

Let us denote by λ and μ the Lamé coefficients appearing in the constitutive equation of a material satisfying Hooke's law:

$$\sigma_{ij} = \lambda\varepsilon_{kk}(u)\delta_{ij} + 2\mu\varepsilon_{ij}(u),$$

(with Einstein's summation convention for repeated indices, and where $\delta_{i,j}$ is the Kronecker delta $\delta_{i,j} = 1$ for $i = j$, $= 0$ for $i \neq j$). Denote the strain tensor by ε :

$$\varepsilon_{ij} = 1/2(\frac{\partial u_i}{\partial x_j} + \frac{\partial u_j}{\partial x_i}).$$

The boundary conditions are given by:

– fixed boundary for Σ (zero displacement);

– a displacement u_a imposed on Γ_a;

– a surface force g imposed on Γ_0.

The f on the right-hand side is the body force density. Let f be an element of $(L^2(\Omega))^3$, u_a of $(H^{1/2}_{0,0}(\mathcal{O}))^3$ and g of $(H^{1/2}_{0,0}(\mathcal{O})')^3$, then the solution of $(\mathcal{PE})$ is well defined in $L^2(0,a;(H^1_0(\mathcal{O}))^3)$ and it depends continuously on the data. For any x_1, we write $\sigma_1(x_1)$ as the first column of the stress tensor defined on Γ_{x_1} and we have that $\sigma_1(x_1) \in (H^{1/2}_{0,0}(\mathcal{O})')^3$.

Other boundary conditions are possible, as for the Laplacian, but these conditions will lead to a more simple calculation for the factorization.

To emphasize the special role of the x_1-coordinate, we will rewrite the elasticity equations in the following form. First, define u' as the projection of u on the plane (x_2, x_3), $u' = (u_2, u_3)^T$. Write div', grad' and Δ', respectively, for the operators of divergence, gradient and the Laplacian in the plane (x_2, x_3). We also write $\mathcal{E}'$ for the elasticity operator in two spatial dimensions:

$$\mathcal{E}'u' = \mu\Delta' u' + (\lambda + \mu)\text{grad}' \,\text{div}'\, u'.$$

Let E and E' be the extension operators to $\mathbb{R}^3$ defined by:

$$Eu_1 = (u_1, 0, 0)^T \qquad E'u' = (0, u_2, u_3)^T.$$

The elasticity equations for deformations can now be rewritten with an emphasis on x_1:

$$\frac{\partial \sigma_1}{\partial x_1} = -f - E\,\text{div}'\,\sigma_1 - \lambda E'\text{grad}'\frac{\partial u_1}{\partial x_1} - \mathcal{E}'u'. \qquad [6.48]$$

Now, define the matrix N and the matrix of operators R by:

$$N = \begin{pmatrix} \dfrac{1}{\lambda + 2\mu} & 0 & 0 \\ 0 & \dfrac{1}{\mu} & 0 \\ 0 & 0 & \dfrac{1}{\mu} \end{pmatrix}, \; R = \begin{pmatrix} 0 & \dfrac{\lambda}{\lambda + 2\mu}\dfrac{\partial}{\partial x_2} & \dfrac{\lambda}{\lambda + 2\mu}\dfrac{\partial}{\partial x_3} \\ \dfrac{\partial}{\partial x_2} & 0 & 0 \\ \dfrac{\partial}{\partial x_3} & 0 & 0 \end{pmatrix}.$$

This allows us to rewrite Hooke's elasticity law for the component σ_1 of the stress tensor:

$$\frac{\partial u}{\partial x_1} = N\sigma_1 - Ru. \qquad [6.49]$$

Substituting this into [6.48], we find that:

$$\frac{\partial \sigma_1}{\partial x_1} = -f - E\,\text{div}'\,\sigma_1 - \lambda E'\text{grad}' e_1^T N\sigma_1 + \lambda E'\text{grad}' e_1^T Ru - \mathcal{E}'u', \qquad [6.50]$$

where e_1 is the unit vector in the x_1-direction. Noting that:

$$-E\text{div}' - \lambda E'\text{grad}'(e_1^T N) = R^\star,$$

and that:

$$\lambda E' \text{grad}' e_1^T R = \frac{\lambda^2}{\lambda + 2\mu} E' \text{grad}' \text{div}',$$

we find that:

$$\frac{\partial \sigma_1}{\partial x_1} - R^\star \sigma_1 - \frac{\lambda^2}{\lambda + 2\mu} E' \text{grad}' \text{div}' u' + \mathcal{E}' u' = -f.$$

Define $\mathcal{M}$ as the matrix:

$$\mathcal{M} = \begin{pmatrix} 0 & 0 \\ 0 & -\mathcal{E}'_P \end{pmatrix},$$

where $\mathcal{E}'_P$ is in turn defined by:

$$\mathcal{E}'_P = \mathcal{E}' - \frac{\lambda^2}{\lambda + 2\mu} \text{grad}' \text{div}' = \mu \Delta' + \mu \frac{3\lambda + 2\mu}{\lambda + 2\mu} \text{grad}' \text{div}'.$$

This yields the equation for σ_1:

$$\frac{\partial \sigma_1}{\partial x_1} - R^\star \sigma_1 = -f + \mathcal{M} u. \quad [6.51]$$

System of equations [6.51], [6.49] with the boundary conditions:

$$-\sigma_1|_{\Gamma_0} = g, \qquad u|_{\Gamma_a} = u_a \quad [6.52]$$

may be viewed as a coupled system of first-order equations in x_1, with forward and backward initial value problems, respectively, for the functions $u(x_1)$ and $\sigma_1(x_1)$. In the following section, we wish to show that this system can be decoupled by applying an invariant embedding method.

REMARK 6.7.– The symmetry of the elasticity system for the deformation tensor ε_{ij} is not preserved in this formulation, due to the special role of the x_1-coordinate.

REMARK 6.8.– The operator $\mathcal{E}'_P$ defined earlier is in fact the operator acting on the displacements (u_2, u_3) with the assumption of plane stress:

$$\sigma_{11} = \sigma_{12} = \sigma_{13} = 0.$$

6.3.2. *Derivation of the decoupled system*

We once again apply the invariant embedding technique of R. Bellman ([BEL 57a]) as in section 1.2. We embed problem $(\mathcal{PE})$ into the family of similar problems $(\mathcal{PE}_{s,h})$ defined on $\Omega_s =]0, s[\times \mathcal{O}$ for $s \in]0, a]$. For each problem, we impose a Dirichlet boundary condition $u|_{\Gamma_s} = h, h \in (H^{1/2}_{0,0}(\mathcal{O}))^3$, where Γ_s is the mobile boundary dependent on the parameter s, $\Gamma_s = \{s\} \times \mathcal{O}$. The piece of data h is therefore an imposed displacement of the boundary Γ_s:

$$(\mathcal{PE}_{s,h}) \begin{cases} -\mu\Delta u - (\lambda + \mu)\mathrm{grad}\,\mathrm{div}\, u = f \quad \text{in } \Omega_s, \\ u|_\Sigma = 0, \\ u|_{\Gamma_s} = h, \\ -\sigma_1|_{\Gamma_0} = g. \end{cases}$$

Problem $(\mathcal{PE})$ corresponds to $(\mathcal{PE}_{s,h})$ for $s = a, h = u_a$. Then, the trace $\sigma_1(s)$ of σ on Γ_s is well defined in $(H^{1/2}_{0,0}(\mathcal{O})')^3$, and by a linearity argument there exists an operator P in $\mathcal{L}((H^{1/2}_{0,0}(\mathcal{O}))^3; (H^{1/2}_{0,0}(\mathcal{O})')^3)$, such that:

$$\sigma_1(s) = P(s)h + w(s). \qquad [6.53]$$

The operator P relates the displacement to the force in the homogeneous problem: it represents a generalized global stiffness of the domain Ω_s. Relation [6.53] must be satisfied for any solution of $(\mathcal{P}_{s,h})$, so that:

$$\sigma_1(x_1) = P(x_1)u(x_1) + w(x_1). \qquad [6.54]$$

We will formally derive an equation for $P(x_1)$ and $w(x_1)$. Justification could possibly be given as in Chapter 2 using a Galerkin method or based on the formulation as a control problem given in section 6.3.3 in a similar way to that followed in Chapter 4, or alternatively working from the framework of Hilbert–Schmidt operators as in [HEN 08]. By differentiating [6.54] w.r.t. x, we obtain:

$$\frac{\mathrm{d}\sigma_1}{\mathrm{d}x_1} = \frac{\mathrm{d}P}{\mathrm{d}x_1}u + P\frac{\mathrm{d}u}{\mathrm{d}x_1} + \frac{\mathrm{d}w}{\mathrm{d}x_1}. \qquad [6.55]$$

From [6.49] and [6.54], we deduce that:

$$\frac{\mathrm{d}u}{\mathrm{d}x_1} = (NP - R)u + Nw, \qquad [6.56]$$

and substituting into [6.55]:

$$\frac{\mathrm{d}\sigma_1}{\mathrm{d}x_1} = (\frac{\mathrm{d}P}{\mathrm{d}x_1} + PNP - PR)u + \frac{\mathrm{d}w}{\mathrm{d}x_1} + PNw. \qquad [6.57]$$

Now, substituting [6.54] into [6.51], we obtain:

$$\frac{\mathrm{d}\sigma_1}{\mathrm{d}x_1} = -f + R^\star(Pu + w) + \mathcal{M}u, \qquad [6.58]$$

then from [6.57], [6.58], we have that:

$$(\frac{\mathrm{d}P}{\mathrm{d}x_1} + PNP - PR - R^\star P - \mathcal{M})u + \frac{\mathrm{d}w}{\mathrm{d}x_1} - R^\star w + PNw + f = 0. \qquad [6.59]$$

As [6.59] is identically true in u, we deduce a Riccati equation for the operator P:

$$\frac{\mathrm{d}P}{\mathrm{d}x_1} + PNP - PR - R^\star P = \mathcal{M}, \qquad [6.60]$$

and an equation for w:

$$\frac{\mathrm{d}w}{\mathrm{d}x_1} - R^\star w + PNw + f = 0. \qquad [6.61]$$

Relation [6.54] evaluated on Γ_0, together with the boundary condition [6.52], yields:

$$w(0) = -g, \qquad P(0) = 0.$$

The decoupled system can now be summarized as:

$$\begin{cases} \frac{\mathrm{d}P}{\mathrm{d}x_1} + PNP - PR - R^\star P = \mathcal{M}, & P(0) = 0, \\ \frac{\mathrm{d}w}{\mathrm{d}x_1} + (PN - R^\star)w = -f, & w(0) = -g, \\ -\frac{\mathrm{d}u}{\mathrm{d}x_1} + (NP - R)u = -Nw, & u(a) = u_a. \end{cases} \qquad [6.62]$$

Note that P is an operator dependent on x_1 acting on functions of x_2, x_3 and satisfying a Riccati equation. System [6.62] is decoupled as the first two equations can be integrated w.r.t. x_1 from 0 to a, which gives P and w, and then u can be obtained by backward integration of the third equation. Formally, we have factorized $-\mathcal{E}u = f$ into:

$$\left(-\frac{\mathrm{d}}{\mathrm{d}x_1} - (PN - R^\star)\right) N^{-1} \left(\frac{\mathrm{d}}{\mathrm{d}x_1} - (NP - R)\right) u = f. \qquad [6.63]$$

It is clear that if P satisfies the Riccati equation in [6.62], then so does its adjoint. By uniqueness of the solution of this equation, P is self-adjoint. Thus, the two factors of [6.63] are mutually adjoint.

As in Chapter 5, we can apply the method of invariant embedding to a discretized version of problem $(\mathcal{PE})$ and obtain the block Gaussian factorization of the discretized problem.

6.3.3. *Associated control problem*

As in Chapter 4, we will show that the elasticity problem $(\mathcal{PE})$ may be viewed as an optimal control problem in which x_1 plays the role of time. The decoupling process resulting in [6.62] in fact corresponds to the derivation of the closed-loop optimal control, i.e. as a function of the present system state. Consider the system with state $u(x_1)$, satisfying the state equation:

$$\frac{\partial u}{\partial x_1} + Ru = N\sigma_1, \qquad u(a) = u_a, \qquad [6.64]$$

where σ_1 is the control variable. Define the "target state":

$$\varphi(x_1) = (0, \varphi_2(x_1), \varphi_3(x_1))^T,$$

for all x_1 as the solution of the 2D elasticity problem with the assumption of plane stress:

$$-\mathcal{E}'_P\varphi(x_1) = -\mu\Delta'\varphi(x_1) - \mu\frac{3\lambda + 2\mu}{\lambda + 2\mu}\mathrm{grad}'\mathrm{div}'\varphi(x_1) = f(x_1)', \quad \text{in } \mathcal{O} \qquad [6.65]$$

$$\varphi(x_1)|_{\partial\mathcal{O}} = 0, \qquad [6.66]$$

with:

$$f(x_1)' = (0, f_2(x_1), f_3(x_1))^T.$$

Next, we define an observation operator C taking values in $\mathbb{R}^3$ by:

$$Cu' = \left(\sqrt{\mu\frac{3\lambda+2\mu}{\lambda+2\mu}}(\frac{\partial u_2}{\partial x_2} + \frac{\partial u_3}{\partial x_3}), \sqrt{\mu}\frac{\partial u_2}{\partial x_2}, \sqrt{\mu}\frac{\partial u_3}{\partial x_3}\right)^T.$$

Now, let J be the cost function:

$$J(\sigma_1) = \frac{1}{2}\int_0^a \int_{\mathcal{O}} \left\|Cu'(\sigma_1; x_1) - C\varphi(x_1)\right\|^2 \,\mathrm{d}x_1\,\mathrm{d}x_2\,\mathrm{d}x_3$$
$$+\frac{1}{2}N\int_\Omega \|\sigma_1\|^2 \,\mathrm{d}x - \int_\Omega u_1 f_1 \,\mathrm{d}x - \int_{\Gamma_0} (u(0), g)\,\mathrm{d}x_2\,\mathrm{d}x_3.$$

The first term of J forces the component u' of the solution u to be close to φ, which is a solution of the 2D plane stress problem [6.65], relative to the observation operator C. Now, the second term of J may be viewed as a regularization term. The existence and uniqueness of the optimal control for the minimization problem of J subject to the constraint [6.64] are guaranteed by coercivity of the quadratic part of J.

Consider the elastic energy functional $\tilde{J}(u)$:

$$\tilde{J}(u) = \frac{1}{2}\int_\Omega \sum_{ij} \sigma_{ij}(u)\varepsilon_{ij}(u)\,\mathrm{d}x - \int_\Omega (f, u)\,\mathrm{d}x - \int_{\Gamma_0} (u(0), g)\,\mathrm{d}x_2\,\mathrm{d}x_3, \qquad [6.67]$$

with the boundary conditions of $(\mathcal{PE})$ (assuming $u_a = 0$ for simplicity). Then, a direct calculation shows that J and $\tilde{J}$ have equal quadratic and linear parts in u and σ, and so only differ by a constant term. This shows the equivalence of the formulation as an optimal control problem and the elasticity system $(\mathcal{PE})$. The following theorem establishes this same result using a different approach.

THEOREM 6.1.– The optimal solution $(\sigma_1^\star, u(\sigma_1^\star))$ of the optimal control problem defined by the minimization of J subject to the constraint [6.64] satisfies system [6.49], [6.51] with the boundary conditions [6.52], or, in other words, u is a solution of $(\mathcal{PE})$. The expression of the optimal feedback of this problem is given by the third equation of [6.62].

PROOF.– Following the method of [LIO 68a], we define the adjoint state p for this optimal control problem by:

$$-\frac{\mathrm{d}p}{\mathrm{d}x_1} + R^\star p = \mathcal{M}(u' - \varphi) - Ef_1, \qquad p(0) = g \text{ on } \mathcal{O}. \qquad [6.68]$$

Using this definition, the optimality condition for $\sigma_1^\star$ is:

$$Np^\star + N\sigma_1^\star = 0, \quad \Rightarrow p^\star = -\sigma_1^\star. \qquad [6.69]$$

Then, by the definition [6.65] of φ, $\sigma_1^\star$ satisfies [6.51] and [6.52]. Now, the optimality system of the optimal control problem is precisely [6.49], [6.51] with the boundary conditions [6.52]. The calculation in the previous section is therefore exactly the calculation that allows the optimal control to be obtained in the form of feedback by decoupling the optimality system. □

6.4. Problems of order higher than 2

In this section, we will explore the decoupling of operators of order higher than 2. We will limit ourselves to operators of order 4, considering the bilaplacian as a representative example. We will continue to consider the geometrically simple situation of the cylinder. With fourth-order operators, the combinatorics of the boundary conditions increase in complexity, and we will only consider two examples in order to demonstrate that the factorization obtained for the Laplacian may be naturally generalized, but that unstable factorizations are also possible.

6.4.1. *A factorization of the bilaplacian*

In the cylinder Ω described in section 1.1, we consider the problem:

$$(\mathcal{PB}) \begin{cases} \Delta^2 u = f \quad \text{in } \Omega, \\ u|_\Sigma = 0, \quad \dfrac{\partial u}{\partial \nu}|_\Sigma = 0, \\ -\dfrac{\partial u}{\partial x}|_{\Gamma_0} = u_0, \quad -\dfrac{\partial \Delta u}{\partial x}|_{\Gamma_0} = v_0 \\ u|_{\Gamma_a} = u_1 \quad \Delta u|_{\Gamma_a} = v_1. \end{cases}$$

We assume that the data have the following regularities:

$$f \in L^2(\Omega), \quad u_1 \in H^{3/2}_{0,0}(\mathcal{O}), \quad u_0 \in H^{1/2}_{0,0}(\mathcal{O}),$$
$$v_1 \in H^{1/2}_{0,0}(\mathcal{O})', \quad v_0 \in H^{3/2}_{0,0}(\mathcal{O})', \qquad [6.70]$$

where the space $H^{3/2}_{0,0}(\mathcal{O})$ is defined as the interpolation space of order $1/2$ between the spaces $H^1_0(\mathcal{O})$ and $H^2_0(\mathcal{O})$. Again, we define the following spaces as "evolution spaces" with respect to the variable x, which are useful for finding a solution and for invariant embedding.

DEFINITION 6.2.–

$$\check{X} = L^2(0,a;H^2_0(\mathcal{O})) \cap H^1(0,a;H^1_0(\mathcal{O})) \cap H^2(0,a;L^2(\mathcal{O})),$$
$$\check{X}_0 = \check{X} \cap \{u \in \check{X} : \ -\frac{\partial u}{\partial x}|_{\Gamma_0} = 0, \ u|_{\Gamma_a} = 0\},$$
$$\check{X}_1 = \check{X} \cap \{u \in \check{X} : \ u|_{\Gamma_a} = 0, \frac{\partial u}{\partial x}|_{\Gamma_a} = 0\}.$$

Problem $(\mathcal{PB})$ has the following variational formulation:

$$a(u,v) = (\Delta u, \Delta v) = (f,v) + (v_1, \frac{\partial v}{\partial x}|_{\Gamma_a}) - (v_0, v|_{\Gamma_0}) \quad \forall v \in \check{X}_0. \quad [6.71]$$

By using liftings of u_0 and u_1 and applying the Lax–Milgram theorem, it can be shown that $(\mathcal{PB})$ is well-posed and has a solution in $\check{X}$, with $-\frac{\partial u}{\partial x}|_{\Gamma_0} = u_0$ and $u|_{\Gamma_a} = u_1$. We can then implement the method of invariant embedding from section 1.2: we consider the family of problems $(\mathcal{PB}_{s,h,k})$ defined on Ω_s by:

$$(\mathcal{PB}_{s,h,k}) \begin{cases} \Delta^2 u_s = f \quad \text{in } \Omega_s, \\ u_s|_\Sigma = 0, \quad \frac{\partial u_s}{\partial \nu}|_\Sigma = 0, \\ -\frac{\partial u_s}{\partial x}|_{\Gamma_0} = u_0, \qquad -\frac{\partial \Delta u_s}{\partial x}|_{\Gamma_0} = v_0, \\ u_s|_{\Gamma_s} = h, \quad \Delta u_s|_{\Gamma_s} = k. \end{cases}$$

Consider the operator that sends the data $u_s|_{\Gamma_s}$, $\Delta u_s|_{\Gamma_s}$ to $\frac{\partial u_s}{\partial x}|_{\Gamma_s}$ and $\frac{\partial \Delta u_s}{\partial x}|_{\Gamma_s}$:

$$\begin{pmatrix} \frac{\partial \Delta u}{\partial x} \\ \frac{\partial u}{\partial x} \end{pmatrix} = \begin{pmatrix} P & Q \\ S & R \end{pmatrix} \begin{pmatrix} \Delta u \\ u \end{pmatrix} + \begin{pmatrix} r \\ s \end{pmatrix}. \qquad [6.72]$$

Differentiating the first line of [6.72] with respect to x:

$$\begin{aligned} \frac{\partial^2 \Delta u}{\partial x^2} =& \frac{\mathrm{d}}{\mathrm{d}x}(Qu + P\Delta u + r) \\ =& (\frac{\mathrm{d}Q}{\mathrm{d}x} + QR + PQ)u + (\frac{\mathrm{d}P}{\mathrm{d}x} + PS + P^2)\Delta u \\ &+ \frac{\mathrm{d}r}{\mathrm{d}x} + Qs + Pr. \end{aligned} \qquad [6.73]$$

Furthermore, from the definition of the bilaplacian and of $(\mathcal{PB})$:

$$\frac{\partial^2 \Delta u}{\partial x^2} = f - \Delta_y \Delta u. \qquad [6.74]$$

Differentiating the second line of [6.72]:

$$\begin{aligned} \frac{\partial^2 u}{\partial x^2} =& \Delta u - \Delta_y u \\ =& (\frac{\mathrm{d}R}{\mathrm{d}x} + R^2 + SQ)u + (\frac{\mathrm{d}S}{\mathrm{d}x} + RS + SP)\Delta u + \frac{\mathrm{d}s}{\mathrm{d}x} + Sr + Rs. \end{aligned} \qquad [6.75]$$

Setting:

$$\mathfrak{P} = \begin{pmatrix} P & Q \\ S & R \end{pmatrix}, \qquad [6.76]$$

we obtain the Riccati equation satisfied by this operator:

$$\frac{\mathrm{d}\mathfrak{P}}{\mathrm{d}x} + \mathfrak{P}^2 = \begin{pmatrix} -\Delta_y & 0 \\ I & -\Delta_y \end{pmatrix}. \qquad [6.77]$$

It can easily be checked that the solution of the equation has the property that $Q(x) = 0$ for all x. It is therefore of the form:

$$\mathfrak{P} = \begin{pmatrix} P & 0 \\ S & R \end{pmatrix}. \qquad [6.78]$$

This result is not obvious *a priori*, because we cannot deduce a problem for the unknown Δu from $(\mathcal{PB})$, as the boundary conditions on Σ do not affect Δu (unlike the situation that we will encounter in Chapter 8). The operators $P \in \mathcal{L}((H^{1/2}_{0,0}(\mathcal{O}))', (H^{3/2}_{0,0}(\mathcal{O}))')$ and $R \in \mathcal{L}(H^{3/2}_{0,0}(\mathcal{O})); H^{1/2}_{0,0}(\mathcal{O}))$ satisfy:

$$\frac{\mathrm{d}P}{\mathrm{d}x} + P^2 = -\Delta_y, \qquad [6.79]$$

and, therefore, both are positive for $x > 0$, as shown in section 1.2. The matrix of operators $\mathfrak{P}$, which is block triangular, is therefore stable. The residual satisfies:

$$\frac{\mathrm{d}}{\mathrm{d}x}\begin{pmatrix} r \\ s \end{pmatrix} + \mathfrak{P}\begin{pmatrix} r \\ s \end{pmatrix} = \begin{pmatrix} f \\ 0 \end{pmatrix}. \qquad [6.80]$$

The initial conditions are:

$$\mathfrak{P}(0) = 0, \qquad r(0) = -v_0, \quad s(0) = -u_0. \qquad [6.81]$$

6.4.2. *Another (unstable) factorization of the bilaplacian*

We now consider the following boundary value problem for the bilaplacian with a Dirichlet condition on Γ_a:

$$(\mathcal{PB}_2)\begin{cases} \Delta^2 u = f \quad \text{in } \Omega, \\ u|_\Sigma = 0, \quad \dfrac{\partial u}{\partial \nu}|_\Sigma = 0, \\ -\Delta u|_{\Gamma_0} + \alpha \dfrac{\partial u}{\partial x}|_{\Gamma_0} = u_0, \quad -\dfrac{\partial \Delta u}{\partial x}|_{\Gamma_0} - \beta u|_{\Gamma_0} = v_0, \quad \alpha > 0, \beta > 0, \\ u|_{\Gamma_a} = u_1 \quad \dfrac{\partial u}{\partial x}|_{\Gamma_a} = v_1. \end{cases}$$

We can show the existence and uniqueness of the solution of this problem in a variational context. Given a test function v in $\check{X}_1$ from definition 6.2, the Green's formula implies that:

$$(\Delta^2 u, v) = (\Delta u, \Delta v) - (\frac{\partial \Delta u}{\partial x}, v)|_{\Gamma_0} + (\Delta u, \frac{\partial v}{\partial x})|_{\Gamma_0}. \qquad [6.82]$$

Hence, starting from the bilinear form:

$$a(u, v) = (\Delta u, \Delta v) + \alpha(\frac{\partial u}{\partial x}, \frac{\partial v}{\partial x})_{\Gamma_0} + \beta(u, v)_{\Gamma_0},$$

we can show the existence of a unique $u \in \check{X}$, such that:

$$a(u, v) = (f, v) - (v_0, v)|_{\Gamma_0} + (u_0, \frac{\partial v}{\partial x})|_{\Gamma_0} \quad \forall v \in \check{X}_1$$

$$\text{with } u|_{\Gamma_a} = u_1 \quad \frac{\partial u}{\partial x}|_{\Gamma_a} = v_1. \qquad [6.83]$$

As before, we consider the family of similar problems defined on the sub-cylinder Ω_s with the Dirichlet conditions u and $\frac{\partial u}{\partial x}$ on Γ_s. Consider the mapping that sends these Dirichlet data to $\frac{\partial^2 u}{\partial x^2}$ and $\frac{\partial^3 u}{\partial x^3}$ on this same section:

$$\begin{pmatrix} \frac{\partial^2 u}{\partial x^2} \\ \frac{\partial^3 u}{\partial x^3} \end{pmatrix} = \begin{pmatrix} P & Q \\ R & S \end{pmatrix} \begin{pmatrix} u \\ \frac{\partial u}{\partial x} \end{pmatrix} + \begin{pmatrix} r \\ t \end{pmatrix}. \qquad [6.84]$$

We differentiate the first line of [6.84] w.r.t. x, and identify with the second line. As the Dirichlet data were arbitrary, we deduce that:

$$\begin{aligned} &\frac{\mathrm{d}P}{\mathrm{d}x} + QP = R, \\ &\frac{\mathrm{d}Q}{\mathrm{d}x} + Q^2 + P = S, \\ &\frac{\mathrm{d}r}{\mathrm{d}x} + Qr = t. \end{aligned} \qquad [6.85]$$

Similarly, differentiating the second line of [6.84], we obtain $\frac{\partial^4 u}{\partial x^4}$, which can be estimated using the equation satisfied by u:

$$\frac{\partial u}{\partial x} = f - 2\Delta_y \frac{\partial^2 u}{\partial x^2} - \Delta_y^2 u. \qquad [6.86]$$

We deduce that:

$$\begin{aligned}
&\frac{\mathrm{d}R}{\mathrm{d}x} + SP = -\Delta_y^2 - 2\Delta_y P,\\
&\frac{\mathrm{d}S}{\mathrm{d}x} + SQ + R = S,\\
&\frac{\mathrm{d}t}{\mathrm{d}x} + Sr = f - 2\Delta_y r.
\end{aligned} \qquad [6.87]$$

By setting:

$$\mathfrak{P} = \begin{pmatrix} P & Q \\ R & S \end{pmatrix},$$

this operator satisfies the Riccati equation:

$$\begin{aligned}
\frac{\mathrm{d}\mathfrak{P}}{\mathrm{d}x} + \mathfrak{P}\begin{pmatrix} 0 & 0 \\ I & 0 \end{pmatrix}\mathfrak{P} + \begin{pmatrix} 0 & -I \\ 2\Delta_y & 0 \end{pmatrix}\mathfrak{P} + \mathfrak{P}\begin{pmatrix} 0 & I \\ 0 & 0 \end{pmatrix}\\
= \begin{pmatrix} 0 & 0 \\ -\Delta_y^2 & 0 \end{pmatrix}.
\end{aligned} \qquad [6.88]$$

We obtain the initial conditions on Γ_0 from the boundary conditions of $(\mathcal{PB}_2)$ and [6.84] at $x = 0$:

$$\begin{aligned}
&P(0) = -\Delta_y; \quad Q(0) = \alpha I; \quad r(0) = -u_0;\\
&\qquad R(0) = -\beta I; \quad S(0) = -\Delta_y; \quad t(0) = -v_0.
\end{aligned} \qquad [6.89]$$

The residual satisfies the equation:

$$\frac{\mathrm{d}}{\mathrm{d}x}\begin{pmatrix} r \\ t \end{pmatrix} + \begin{pmatrix} Q & -I \\ S + 2\Delta_y & 0 \end{pmatrix}\begin{pmatrix} r \\ t \end{pmatrix} = \begin{pmatrix} 0 \\ f \end{pmatrix}. \qquad [6.90]$$

The solution of $(\mathcal{PB}_2)$ is obtained by solving the initial value system in the backward direction:

$$\frac{\mathrm{d}}{\mathrm{d}x}\begin{pmatrix} u \\ u' \end{pmatrix} = \begin{pmatrix} 0 & I \\ P & Q \end{pmatrix}\begin{pmatrix} u \\ u' \end{pmatrix} + \begin{pmatrix} 0 \\ r \end{pmatrix}. \qquad [6.91]$$

REMARK 6.9.– Note that this factorization is not self-adjoint. Furthermore, the initial value problems [6.91], [6.90] arising from the factorization are not necessarily well-posed, as the corresponding dynamics operators are not positive. In particular, in the case where $\alpha = \beta = 0$, the Riccati equation [6.88] admits the constant solution:

$$\mathfrak{P} = \begin{pmatrix} -\Delta_y & 0 \\ 0 & -\Delta_y \end{pmatrix}. \qquad [6.92]$$

In this case, using [6.84], the residual r is the Laplacian of the solution:

$$r(x) = \Delta u|_{\Gamma_x}; \quad t(x) = \frac{\mathrm{d}}{\mathrm{d}x}\Delta u|_{\Gamma_x}. \qquad [6.93]$$

Problems [6.91], [6.90] are Cauchy problems for the Laplacian that are well known for being unstable [HAD 10]. Consider further a discretization of problem $(\mathcal{PB}_2)$ and apply the block Gaussian factorization to the resulting matrix with respect to the previous factorization as explained in Chapter 5. Without a strategy of row and/or column permutation (see section 3.5), this Gaussian factorization will be numerically unstable.

6.5. Stokes problems

The factorization method was implemented for the Stokes equation by F. Jday in his thesis [JDA 12]. Consider a Stokes problem in the cylinder in $\mathbb{R}^3$, $\Omega =]0, a[\times\mathcal{O}$. Denote by $u \in \mathbb{R}^3$ the velocity of the fluid, p the pressure and ν the viscosity. The function $f \in (L^2(\Omega))^3$ on the right-hand side is a field of distributed forces. As before, x is the coordinate along the axis of the cylinder and is taken as an index for vector components in this direction. Similarly, y represents the coordinates on the section $\mathcal{O}$ and is taken as a multi-index accordingly. The Stokes problem is then written as follows:

$$(\mathcal{PS}) \quad \begin{cases} -\nu\Delta u + \nabla p = f & \text{in } \Omega, \\ \text{div} u = 0 & \text{in } \Omega, \\ u = 0 & \text{on } \partial\Omega \setminus \Gamma_0, \\ \nu \dfrac{\partial u_y}{\partial x} = 0 & \text{on } \Gamma_0, \\ \nu(\dfrac{\partial u_x}{\partial x} + p) = 0 & \text{on } \Gamma_0. \end{cases} \qquad [6.94]$$

To avoid a singularity due to a Dirichlet condition on Γ_0, such as that encountered in section 3.2.2, we replace it with a Neumann-type condition, also known as a stress condition for the Stokes problem [BOY 06]. The cited work examines both types of boundary condition and shows that, in the first case (Dirichlet), the velocity is uniquely determined, but the pressure is only known up to an additive constant; in the second case (Neumann), the pressure is unique but the velocity is only known if an additional condition is imposed, for example, setting the average to zero. Here, the mixed conditions allow us to uniquely determine both velocity and pressure. Let X_{div} be the space:

$$X_{\text{div}} = \{u \in (H^1(\Omega))^3 | \text{div} u = 0 \text{ in } \Omega, u|_{\partial\Omega\setminus\Gamma_0} = 0\}. \qquad [6.95]$$

We obtain the variational formulation of the problem by taking a test function $v \in X_{\text{div}}$ and applying Green's formula:

$$\begin{aligned} \int_\Omega (-\nu\Delta u + \nabla p) v \, \mathrm{d}x \, \mathrm{d}y &= \int_\Omega \nu \nabla u \nabla v \, \mathrm{d}x \, \mathrm{d}y - \nu \int_{\Gamma_0} \frac{\partial u}{\partial n} v \, \mathrm{d}y \\ &\quad + \int_{\Gamma_0} p v_x \, \mathrm{d}y \\ &= \int_\Omega \nu \nabla u \nabla v \, \mathrm{d}x \, \mathrm{d}y + \nu \int_{\Gamma_0} \frac{\partial u_y}{\partial x} v_y \, \mathrm{d}y \qquad [6.96] \\ &\quad + \nu \int_{\Gamma_0} \frac{\partial u_x}{\partial x} v_x \, \mathrm{d}y + \int_{\Gamma_0} p v_x \, \mathrm{d}y \\ &= \int_\Omega \nu \nabla u \nabla v \, \mathrm{d}x \, \mathrm{d}y. \end{aligned}$$

By applying the Lax–Milgram theorem to this variational formulation, we can show the existence and uniqueness of a weak solution of [6.94]. To simplify this calculation, we will assume in the following that $\nu = 1$.

We will now show that we can once again apply the method of invariant embedding to obtain the factorization of the Stokes problem. We embed problem [6.94] into a family of similar problems defined on $\Omega_s =]0, s[\times\mathcal{O}$, for $s \in]0, a[$. For each problem, we impose a Dirichlet boundary condition $h = (h_x, h_y) \in (H_{0,0}^{1/2}(\mathcal{O}))^3$ on the face Γ_s. Consider now the solution of [6.94]. Its restriction to $\Omega \setminus \Omega_s$ implies that the global flux through Γ_s must vanish due to incompressibility and the other boundary conditions. However, for purposes of invariant embedding, we can choose h such that the corresponding solution of $(\mathcal{PS}_{s,h})$ (which we shall identify more precisely in [6.97]) does not necessarily satisfy this condition. Indeed, from the stress condition on Γ_0 which does not fix the total flux, we obtain a well-posed problem on Ω_s:

$$(\mathcal{PS}_{s,h}) \quad \begin{cases} -\Delta u^s + \nabla p^s = f & \text{in} \quad \Omega_s, \\ \text{div} u^s = 0 & \text{in} \quad \Omega_s, \\ u^s = 0 & \text{on} \quad \Sigma_s, \\ \dfrac{\partial u_y^s}{\partial x} = 0 & \text{on} \quad \Gamma_0, \\ \dfrac{\partial u_x^s}{\partial x} + p^s = 0 & \text{on} \quad \Gamma_0, \\ u^s = h \in (H_{0,0}^{1/2}(\mathcal{O}))^3 & \text{on} \quad \Gamma_s. \end{cases} \qquad [6.97]$$

Note that here, as in section 3.2.4, the set of boundary conditions considered on the mobile section for this family of problems is not the same as that of the initial problem. In this case, it is larger.

As before, we separate u^s and p^s into parts that are linearly dependent on h, γ_s and p_l^s, and parts that are independent of h, β^s and p_0^s. The pressure $p_0^s(s)$ is the pressure on the section Γ_s given vanishing Dirichlet conditions on Γ_s for problem [6.97]. We have that:

$$\begin{cases} -\Delta \gamma^s + \nabla p_l^s = 0 & \text{in} \quad \Omega_s, \\ \text{div} \gamma^s = 0 & \text{in} \quad \Omega_s, \\ \gamma^s = 0 & \text{on} \quad \Sigma_s, \\ \dfrac{\partial \gamma_y^s}{\partial x} = 0 & \text{on} \quad \Gamma_0, \\ \dfrac{\partial \gamma_x^s}{\partial x} + p_l^s = 0 & \text{on} \quad \Gamma_0, \\ \gamma^s = h & \text{on} \quad \Gamma_s, \end{cases} \qquad [6.98]$$

$$\begin{cases} -\Delta\beta^s + \nabla p_a^s = f & \text{in} \quad \Omega_s, \\ \text{div}\beta^s = 0 & \text{in} \quad \Omega_s, \\ \beta^s = 0 & \text{on} \quad \Sigma_s, \\ \dfrac{\partial \beta_y^s}{\partial x} = 0 & \text{on} \quad \Gamma_0, \\ \dfrac{\partial \beta_x^s}{\partial x} + p_0^s = 0 & \text{on} \quad \Gamma_0, \\ \beta^s = 0 & \text{on} \quad \Gamma_s. \end{cases} \qquad [6.99]$$

We therefore have:

$$u^s = \beta^s + \gamma^s, \quad p^s = p_l^s + p_0^s.$$

By reapplying Green's formula [6.96] with test functions that are non-zero on Γ_s, we can define the normal derivative of u^s on Γ_s in $((H_{0,0}^{1/2}(\mathcal{O}))^3)'$. The Dirichlet-to-Neumann mapping $P(s)$ is given by:

$$P(s) : (H_{0,0}^{1/2}(\mathcal{O}))^3 \mapsto ((H_{0,0}^{1/2}(\mathcal{O}))^3)', \\ h \mapsto Ph = \frac{\partial \gamma^s}{\partial x}|_{\Gamma_s}. \qquad [6.100]$$

The residual w is defined by:

$$w(s) = \frac{\partial \beta^s}{\partial x}. \qquad [6.101]$$

We therefore have:

$$\frac{\partial u^s}{\partial x}(x;h)|_{\Gamma_s} = P(s)h + w(s). \qquad [6.102]$$

Taking h to be the trace of the solution of [6.94] on Γ_x and setting $x = s$, we obtain the following identity:

$$\frac{\partial u(x)}{\partial x}|_{\Gamma_x} = P(x)u(x) + w(x). \qquad [6.103]$$

Decomposing the operator P into blocks as a function of the components u_x and u_y of u:

$$P = \begin{pmatrix} P_{x,x} & P_{x,y} \\ P_{y,x} & P_{y,y} \end{pmatrix}. \qquad [6.104]$$

The second equation [6.94] gives:

$$\frac{\partial u_x}{\partial x} = -\mathrm{div}_y u_y, \qquad [6.105]$$

and so P becomes:

$$P = \begin{pmatrix} 0 & -\mathrm{div}_y \\ P_{y,x} & P_{y,y} \end{pmatrix}. \qquad [6.106]$$

Relation [6.103] gives:

$$\frac{\partial u_y(x)}{\partial x}|_{\Gamma_s} = P_{y,y}(x)u_y(x) + P_{y,x}(x)u_x(x) + w_y(x). \qquad [6.107]$$

Differentiating this relation with respect to x and using [6.103]:

$$\begin{aligned}\frac{\partial^2 u_y}{\partial x^2} =& \frac{\mathrm{d}P_{y,y}}{\mathrm{d}x}u_y + P_{y,y}\frac{\partial u_y}{\partial x} + \frac{\mathrm{d}P_{y,x}}{\mathrm{d}x}u_x + P_{y,x}\frac{\partial u_x}{\partial x} + \frac{\partial w_y}{\partial x} \\ =& \frac{\mathrm{d}P_{y,y}}{\mathrm{d}x}u_y + P_{y,y}^2 u_y + P_{y,y}P_{y,x}u_x + P_{y,y}w_y + \frac{\mathrm{d}P_{y,x}}{\mathrm{d}x}u_x \\ &- P_{y,x}\mathrm{div}_y u_y + \frac{\partial w_y}{\partial x}.\end{aligned}$$

The first equation of [6.94] gives:

$$\frac{\partial^2 u_y}{\partial x^2} = -\Delta_y u_y + \nabla_y p - f_y. \qquad [6.108]$$

Then, we have:

$$\begin{aligned}-\Delta_y u_y + \nabla_y p - f_y =& \frac{\mathrm{d}P_{y,y}}{\mathrm{d}x}u_y + P_{y,y}^2 u_y + P_{y,y}P_{y,x}u_x \\ &+ \frac{\mathrm{d}P_{y,x}}{\mathrm{d}x}u_x - P_{y,x}\mathrm{div}_y u_y + P_{y,y}w_y + \frac{\partial w_y}{\partial x}.\end{aligned} \qquad [6.109]$$

Relation [6.107] evaluated on Γ_0 for the linear problem yields:

$$\frac{\partial \gamma_y^s}{\partial x}(0) = P_{y,y}(0)\gamma_y^s(0) + P_{y,x}(0)\gamma_x^s(0),$$

and from the fourth equation of [6.98]:

$$\frac{\partial \gamma_y^s}{\partial x}(0) = 0.$$

Then, by identification between these two relations, we obtain:

$$P_{y,y}(0) = 0, \qquad P_{y,x}(0) = 0. \tag{6.110}$$

And consequently, from relation [6.107] and the fourth equation of [6.94], we deduce that:

$$w_y(0) = 0.$$

We now introduce the mappings $Q(s)$ and $R(s)$ defined by:

$$p(s) = Q(s)h_y + R(s)h_x + p_0(s), \quad \forall h \in (H_{0,0}^{1/2}(\mathcal{O}))^3. \tag{6.111}$$

These mappings are operators defined on the section Γ_s that depend only on s. Returning to [6.109] and substituting the value of p given by [6.111]:

$$\begin{aligned}
&(-\Delta_y + \nabla_y Q - \frac{\mathrm{d}P_{y,y}}{\mathrm{d}x} - P_{y,y}^2 + P_{y,x}\mathrm{div}_y)u_y \\
&\quad + (\nabla_y R - P_{y,y}P_{y,x} - \frac{\mathrm{d}P_{y,x}}{\mathrm{d}x})u_x \\
&\quad - P_{y,y}w_y - \frac{\partial w_y}{\partial x} - f_y + \nabla_y p_0 = 0.
\end{aligned} \tag{6.112}$$

We also have that:

$$-\Delta u_x = -\frac{\partial^2 u_x}{\partial x^2} - \Delta_y u_x, \tag{6.113}$$

and from [6.105]:

$$\begin{aligned} -\Delta u_x &= \frac{\partial}{\partial x}(\mathrm{div}_y u_y) - \Delta_y u_x \\ &= \mathrm{div}_y \frac{\partial u_y}{\partial x} - \Delta_y u_x. \end{aligned} \qquad [6.114]$$

Replacing $\frac{\partial u_y}{\partial x}$ by [6.107]:

$$-\Delta u_x = \mathrm{div}_y (P_{y,y} u_y + P_{y,x} u_x + w_y) - \Delta_y u_x.$$

From the first equation of system [6.94], we have:

$$\frac{\partial p}{\partial x} = \Delta u_x + f_x. \qquad [6.115]$$

Replacing Δu_x by the above expression, we obtain:

$$\frac{\partial p}{\partial x} = -\mathrm{div}_y (P_{y,y} u_y + P_{y,x} u_x + w_y) + \Delta_y u_x + f_x.$$

Identifying this quantity with the derivative w.r.t. x of [6.111], we have that:

$$\frac{\partial}{\partial x}(Q u_y + R u_x + p_0) = -\mathrm{div}_y (P_{y,y} u_y + P_{y,x} u_x + w_y) + \Delta_y u_x + f_x.$$

Replacing $\frac{\partial u_y}{\partial x}$ by [6.107] and $\frac{\partial u_x}{\partial x}$ by $-\mathrm{div}_y u_y$, we find that:

$$\begin{aligned} &\frac{\mathrm{d}Q}{\mathrm{d}x} u_y + Q P_{y,y} u_y + Q P_{y,x} u_x + Q w_y + \frac{\mathrm{d}R}{\mathrm{d}x} u_x - R \mathrm{div}_y u_y \\ &\quad + \frac{\partial p_0}{\partial x} = -\mathrm{div}_y P_{y,y} u_y - \mathrm{div}_y P_{y,x} u_x - \mathrm{div}_y w_y + \Delta_y u_x + f_x, \end{aligned}$$

from which it follows that:

$$\begin{aligned} &(\frac{\mathrm{d}Q}{\mathrm{d}x} + Q P_{y,y} - R \mathrm{div}_y + \mathrm{div}_y P_{y,y}) u_y + \\ &(Q P_{y,x} + \frac{\mathrm{d}R}{\mathrm{d}x} + \mathrm{div}_y P_{y,x} - \Delta_y) u_x + \frac{\partial p_0}{\partial x} + \mathrm{div}_y w_y - f_x = 0. \end{aligned} \qquad [6.116]$$

As h is arbitrary in $(H^{1/2}_{0,0}(\mathcal{O}))^3$, we obtain that:

$$(\frac{\mathrm{d}Q}{\mathrm{d}x} + QP_{y,y} - R\mathrm{div}_y + \mathrm{div}_y P_{y,y})u_y = 0 \quad \forall u_y \in (H^{1/2}_{0,0}(\mathcal{O}))^2, \qquad [6.117]$$

$$(\frac{\mathrm{d}R}{\mathrm{d}x} + QP_{y,x} + \mathrm{div}_y P_{y,x} - \Delta_y)u_x = 0 \quad \forall u_x \in H^{1/2}_{0,0}(\mathcal{O}). \qquad [6.118]$$

Relation [6.111], for $x = 0$, gives:

$$p(0) = Q(0)u_y(0) + R(0)u_x(0) + p_0(0)$$

and the fifth equation of [6.94] gives:

$$p(0) = \mathrm{div}_y u_y(0).$$

Then, by identification between these two relations, we find:

$$Q(0) = \mathrm{div}_y, \quad R(0) = 0, \quad p_0(0) = 0. \qquad [6.119]$$

Regrouping equations [6.107], [6.112], [6.116], [6.111] and [6.105] and the initial conditions introduced in [6.110] and [6.119], we obtain the following two decoupled systems of equations:

$$\begin{cases} \dfrac{\mathrm{d}P_{y,y}}{\mathrm{d}x} + \Delta_y - \nabla_y Q + P^2_{y,y} - P_{y,x}\mathrm{div}_y = 0, & P_{y,y}(0) = 0, \\ \dfrac{\mathrm{d}P_{y,x}}{\mathrm{d}x} - \nabla_y R + P_{y,y}P_{y,x} = 0, & P_{y,x}(0) = 0, \\ \dfrac{\mathrm{d}Q}{\mathrm{d}x} + QP_{y,y} - R\mathrm{div}_y + \mathrm{div}_y P_{y,y} = 0, & Q(0) = \mathrm{div}_y, \\ \dfrac{\mathrm{d}R}{\mathrm{d}x} + QP_{y,x} + \mathrm{div}_y P_{y,x} + \Delta_y = 0, & R(0) = 0, \end{cases} \qquad [6.120]$$

$$\begin{cases} \dfrac{\partial w_y}{\partial x} + P_{y,y}w_y + f_y = 0, & w(0) = 0, \\ \dfrac{\partial p_0}{\partial x} + \mathrm{div}_y w_y - f_x = 0, & p_0(0) = 0, \\ \dfrac{\partial u_x}{\partial x} + \mathrm{div}_y u_y = 0, & u_x(a) = 0, \\ \dfrac{\partial u_y}{\partial x} - P_{y,y}u_y - P_{y,x}u_x - w_y(x) = 0, & u_y(a) = 0, \\ p - Qu_y - Ru_x - p_0 = 0. \end{cases} \qquad [6.121]$$

System [6.120] can be written in the form of a Riccati equation for a matrix of operators. Let:

$$\mathfrak{C} = \begin{pmatrix} 0 & 0 \\ \mathrm{div}_y & 0 \end{pmatrix}, \quad \mathfrak{D} = \begin{pmatrix} 0 & \nabla_y \\ -\mathrm{div}_y & 0 \end{pmatrix}, \quad \mathfrak{E} = \begin{pmatrix} I & 0 \\ 0 & 0 \end{pmatrix}, \quad [6.122]$$

and:

$$\mathfrak{P} = \begin{pmatrix} P_{y,y} & P_{y,x} \\ Q & R \end{pmatrix}. \quad [6.123]$$

Then, system [6.120] may be written as:

$$\frac{\mathrm{d}\mathfrak{P}}{\mathrm{d}x} + \mathfrak{P}\mathfrak{E}\mathfrak{P} - \mathfrak{P}\mathfrak{C} - \mathfrak{D}\mathfrak{P} = -\begin{pmatrix} \Delta_y & 0 \\ 0 & \Delta_y \end{pmatrix}. \quad [6.124]$$

REMARK 6.10.– Note that in this case it was not possible to relate the considered problem to an optimal control problem, and the operator $\mathfrak{P}$ is not self-adjoint.

REMARK 6.11.– The operator $\mathfrak{P}$ has a physical meaning, as it relates the velocity imposed at the boundary Γ_s to the resulting pressure and shear rate on the same boundary.

6.6. Parabolic problems

In this section, we wish to study extensions of the factorization method to parabolic evolution problems. We consider the heat equation in a cylindrical spatial domain, which is the evolution problem analogous to $(\mathcal{P}_0)$ from section 1.1:

$$(\mathcal{PP}_0) \begin{cases} \dfrac{\partial u}{\partial t} - \Delta u = f \quad \text{in } Q_T = \Omega \times]0,T[, \\ u|_{\Sigma\times]0,T[} = 0, \quad -\dfrac{\partial u}{\partial x}|_{\Gamma_0\times]0,T[} = u_0, \quad u|_{\Gamma_a\times]0,T[} = u_1 \\ \text{with the initial condition } u|_{t=0} = u^0 \in L^2(\Omega). \end{cases}$$

This problem has a unique solution in $L^2(0,T;X)$, such that $\dfrac{\partial u}{\partial t} \in L^2(0,T;X')$, where X is defined by definition 1.1.

The method of invariant embedding allows boundary value problems to be transformed into a product of Cauchy problems. Invariant embedding in time is not useful, as problem $(\mathcal{P}\mathcal{P}_0)$ is already a Cauchy problem. We will therefore choose a time interval $]0,t[, 0 < t < T$ and proceed to perform spatial invariant embedding with respect to the variable x along the axis of the cylinder.

As in section 3.1, we will present a Green's formula that, together with a family of test functions that extends the usual variational formulation, allows invariant embedding to be achieved. For functions that are sufficiently regular or zero on Σ, we have Green's formula:

$$\int_{Q_t}\left(\frac{\partial u}{\partial t'} - \Delta u\right) v \,\mathrm{d}x\,\mathrm{d}y\,\mathrm{d}t' = -\int_{Q_t} u\frac{\partial v}{\partial t'}\,\mathrm{d}x\,\mathrm{d}y\,\mathrm{d}t' + \int_{\Omega} u(t)v(t)\,\mathrm{d}x\,\mathrm{d}y$$
$$-\int_{\Omega} u^0 v(0)\,\mathrm{d}x\,\mathrm{d}y + \int_{Q_t} \nabla u \nabla v\,\mathrm{d}x\,\mathrm{d}y\,\mathrm{d}t' - \int_{\Sigma\times]0,t[} \frac{\partial u}{\partial \nu} v\,\mathrm{d}\Sigma\,\mathrm{d}t'$$
$$-\int_{\Gamma_a\times]0,t[} \frac{\partial u}{\partial x} v\,\mathrm{d}y\,\mathrm{d}t' + \int_{\Gamma_0\times]0,t[} \frac{\partial u}{\partial x} v\,\mathrm{d}y\,\mathrm{d}t'. \qquad [6.125]$$

Now, let u be a solution of $(\mathcal{P}\mathcal{P}_0)$. We can extend formula [6.125] to (at least) functions v in $L^2(0,t;H^1(\Omega))$ with $\frac{\mathrm{d}v}{\mathrm{d}t} \in L^2(0,t;H^1(\Omega)')$. Using the regularity of u, this formula allows us to define $\frac{\partial u}{\partial x}$ on Γ_0 and Γ_a in $L^2(0,t;H^{1/2}_{0,0}(\mathcal{O})')$.

For invariant embedding, we consider the spatial domain $\Omega_x = \mathcal{O}\times]0,x[$ and the corresponding time cylinder $Q_{t,x} = \Omega_x\times]0,t[$. We fix an arbitrary Dirichlet condition h on $\Gamma_x\times]0,t[$. For the parabolic problem thus defined, we can calculate the normal derivative of the solution on $\Gamma_x\times]0,t[$, which depends affinely on the data hq. We therefore have the Dirichlet-to-Neumann operator $P_t(x) \in \mathcal{L}(L^2(0,t;H^{1/2}_{0,0}(\mathcal{O})); L^2(0,t;H^{1/2}_{0,0}(\mathcal{O})'))$ and $P_t(x) \in \mathcal{L}(L^2(0,t;H^1_0(\mathcal{O})); L^2(0,t;L^2(\mathcal{O})))$ and a residual function w, such that:

$$\int_{\Gamma_x\times]0,t[} \frac{\partial u}{\partial x} v|_{\Gamma_x\times]0,t[}\,\mathrm{d}y\,\mathrm{d}t' = \int_{\Gamma_x\times]0,t[} P_t(x)hv|_{\Gamma_x\times]0,t[}\,\mathrm{d}y\,\mathrm{d}t' + \int_{\Gamma_x\times]0,t[} wv|_{\Gamma_x\times]0,t[}\,\mathrm{d}y\,\mathrm{d}t'. \quad [6.126]$$

We consider Green's formula [6.125] for the spatio-temporal domain $Q_{t,x}$ with test functions v independent of x and zero on Σ: $v \in L^2(0,t;H_0^1(\mathcal{O}))$, such that $\frac{\partial v}{\partial t'} \in L^2(0,t;H^{-1}(\mathcal{O}))$ and $v(t) = 0$ on Σ. Using the fact that u satisfies $(\mathcal{PP}_0)$ and [6.126], we have that:

$$\begin{aligned}\int_{\Gamma_x\times]0,t[} (P_t(x)u(x) + w(x))\,v\,\mathrm{d}y\,\mathrm{d}t' &= -\int_{Q_{t,x}} fv\,\mathrm{d}x\,\mathrm{d}y\,\mathrm{d}t' \\ &- \int_{Q_{t,x}} u\frac{\partial v}{\partial t'}\,\mathrm{d}x\,\mathrm{d}y\,\mathrm{d}t' - \int_{\Omega_x} u^0 v(0)\,\mathrm{d}x\,\mathrm{d}y \\ &+ \int_{Q_{t,x}} \nabla_y u \nabla_y v\,\mathrm{d}x\,\mathrm{d}y\,\mathrm{d}t' + \int_{\Gamma_0\times]0,t[} \frac{\partial u}{\partial x} v\,\mathrm{d}y\,\mathrm{d}t'. \end{aligned} \quad [6.127]$$

By formally differentiating w.r.t. x, we obtain:

$$\begin{aligned}\int_{\Gamma_x\times]0,t[} \frac{\mathrm{d}}{\mathrm{d}x}(P_t(x)u(x) + w(x))\,v\,\mathrm{d}y\,\mathrm{d}t' &= -\int_{\Gamma_x\times]0,t[} fv\,\mathrm{d}y\,\mathrm{d}t' \\ &- \int_{\Gamma_x\times]0,t[} u\frac{\partial v}{\partial t'}\,\mathrm{d}y\,\mathrm{d}t' - \int_{\Gamma_x} u^0 v(0)\,\mathrm{d}y + \int_{\Gamma_x\times]0,t[} \nabla_y u \nabla_y v\,\mathrm{d}y\,\mathrm{d}t'. \end{aligned} \quad [6.128]$$

The left-hand side gives:

$$\begin{aligned}\int_{\Gamma_x\times]0,t[} \frac{\mathrm{d}}{\mathrm{d}x}(P_t(x)u(x) + w(x))\,v\,\mathrm{d}y\,\mathrm{d}t' &= \\ = \int_{\Gamma_x\times]0,t[} \left(\frac{\mathrm{d}P_t(x)}{\mathrm{d}x} + P_t^2(x)\right) u(x)v\,\mathrm{d}y\,\mathrm{d}t' \\ + \int_{\Gamma_x\times]0,t[} \left(\frac{\mathrm{d}w(x)}{\mathrm{d}x} + P_t(x)w(x)\right) v\,\mathrm{d}y\,\mathrm{d}t'. \end{aligned} \quad [6.129]$$

We substitute this calculation into [6.128], and we appeal to the argument of Chapter 1 regarding arbitrary trajectories (in x), so that we can separate the part linearly dependent on u from the part independent of u:

$$\int_{\Gamma_x\times]0,t[}\left(\frac{\mathrm{d}P_t(x)}{\mathrm{d}x}+P_t^2(x)\right)u(x)v\,\mathrm{d}y\,\mathrm{d}t' = -\int_{\Gamma_x\times]0,t[} u\frac{\partial v}{\partial t'}\,\mathrm{d}y\,\mathrm{d}t' + \int_{\Gamma_x\times]0,t[} \nabla_y u\nabla_y v\,\mathrm{d}y\,\mathrm{d}t'. \quad [6.130]$$

$$\int_{\Gamma_x\times]0,t[}\left(\frac{\mathrm{d}w(x)}{\mathrm{d}x}+P_t(x)w(x)\right)v\,\mathrm{d}y\,\mathrm{d}t = -\int_{\Gamma_x\times]0,t[} fv\,\mathrm{d}y\,\mathrm{d}t' - \int_{\Gamma_x} u^0 v(0)\,\mathrm{d}y. \quad [6.131]$$

The initial conditions for these evolution equations in x result from the boundary condition on $\Gamma_0\times]0,t[$:

$$P_t(0)=0, \quad w(0)=-u_0. \quad [6.132]$$

These equations are complemented by [6.126], which can be viewed as an evolution equation in the direction of decreasing x with an initial condition at $x=a$:

$$\int_{\Gamma_x\times]0,t[}\left(\frac{\mathrm{d}u(x)}{\mathrm{d}x}-P_t(x)u(x)\right)v\,\mathrm{d}y\,\mathrm{d}t = \int_{\Gamma_x\times]0,t[} wv\,\mathrm{d}y\,\mathrm{d}t, \quad u(a)=u_1. \quad [6.133]$$

REMARK 6.12.– Unlike the elliptic case, the Dirichlet-to-Neumann operator defined by [6.126] and satisfying the Riccati equation [6.130] is neither self-adjoint nor positive.

We now wish to explicitly exhibit the effect of the initial condition u^0 on $u(t)$. To achieve this, we proceed as in section 3.3 for the Dirichlet conditions on the lateral boundary Σ for the elliptic problem by isolating in w the part that depends uniquely on u^0. Relation [6.126] becomes:

$$\int_{\Gamma_x\times]0,t[} \frac{\partial u}{\partial x} v|_{\Gamma_x\times]0,t[} \,\mathrm{d}y\,\mathrm{d}t' = \int_{\Gamma_x\times]0,t[} P_t(x) h v|_{\Gamma_x\times]0,t[} \,\mathrm{d}y\,\mathrm{d}t'$$
$$+ \int_0^x \left(T_t(x',x)u^0(x'), v\right)_{L^2(\Gamma_x\times]0,t[)} \,\mathrm{d}x'$$
$$+ \int_{\Gamma_x\times]0,t[} w' v|_{\Gamma_x\times]0,t[} \,\mathrm{d}y\,\mathrm{d}t', \tag{6.134}$$

where $T_t(x',x) \in \mathcal{L}(L^2(\mathcal{O}); L^2(0,t; H^{1/2}_{0,0}(\mathcal{O})'))$ and $T_t(.,.) \in \mathcal{L}(L^2(\Omega); L^2(0,t; L^2(\Omega)))$. Formula [6.129] becomes:

$$\int_{\Gamma_x\times]0,t[} \frac{\mathrm{d}}{\mathrm{d}x}\left(P_t(x)u(x) + \int_0^x T_t(x',x)u^0(x')\,\mathrm{d}x' + w'(x)\right) v\,\mathrm{d}y\,\mathrm{d}t' =$$
$$= \int_{\Gamma_x\times]0,t[} \left(\frac{\mathrm{d}P_t(x)}{\mathrm{d}x} + P_t^2(x)\right) u(x) v\,\mathrm{d}y\,\mathrm{d}t'$$
$$+ \int_{\Gamma_x\times]0,t[} \left(\int_0^x P_t(x)T_t(x,x')u^0(x')\,\mathrm{d}x'\right) v\,\mathrm{d}y\,\mathrm{d}t'$$
$$+ \int_{\Gamma_x\times]0,t[} \left(T_t(x,x)u^0(x) + \int_0^x \frac{\mathrm{d}}{\mathrm{d}x} T_t(x,x')u^0(x')\,\mathrm{d}x'\right) v\,\mathrm{d}y\,\mathrm{d}t'$$
$$+ \int_{\Gamma_x\times]0,t[} \left(\frac{\mathrm{d}w'(x)}{\mathrm{d}x} + P_t(x)w'(x)\right) v\,\mathrm{d}y\,\mathrm{d}t'. \tag{6.135}$$

By identifying this with the right-hand side of [6.128] and using the fact that u, v and also u^0 are arbitrary, as before, we find that P_t satisfies equation [6.130] and T_t satisfies:

$$\frac{\mathrm{d}}{\mathrm{d}x} T_t(x,x') + P_t(x)T_t(x,x') = 0, \tag{6.136}$$

$$\int_{\Gamma_x\times]0,t[} T_t(x,x)u^0(x) v\,\mathrm{d}y\,\mathrm{d}t' = -\int_{\Gamma_x} u^0(x)v(0)\,\mathrm{d}y. \tag{6.137}$$

The residual w' satisfies:

$$\int_{\Gamma_x\times]0,t[} \left(\frac{\mathrm{d}w'(x)}{\mathrm{d}x} + P_t(x)w'(x)\right) v\,\mathrm{d}y\,\mathrm{d}t = -\int_{\Gamma_x\times]0,t[} fv\,\mathrm{d}y\,\mathrm{d}t', \qquad w'(0) = -u_0. \tag{6.138}$$

The solution u of the problem is now given by:

$$\int_{\Gamma_x\times]0,t[} \left(\frac{\mathrm{d}u(x)}{\mathrm{d}x} - P_t(x)u(x)\right) v\,\mathrm{d}y\,\mathrm{d}t = \int_{\Gamma_x\times]0,t[} w'v\,\mathrm{d}y\,\mathrm{d}t + \int_0^x \left(T_t(x',x)u^0(x'),v\right)_{L^2(\Gamma_x\times]0,t[)} \mathrm{d}x', \qquad u(a) = u_1. \tag{6.139}$$

If we assume that P_t and the family $T_t(x',x)$ for $0 \le x' \le x \le a$ are given, we can calculate u by integrating [6.138] for x ranging from 0 to a, and then [6.139] from a to 0. Once u has been calculated, we can take its trace w.r.t. t, $u|_t$. Thus, we obtain a factorization of the evolution operator for the parabolic equation: $u_0 \mapsto u|_t$. We observe that, in this calculation, the values of u_0 are introduced on intervals $]0,x[$ for increasing x, whereas u and thus $u|_t$ are obtained on intervals $]x,a[$ for decreasing x.

REMARK 6.13.– The above approach can also be used as a discretization scheme for the parabolic problem $(\mathcal{PP}_0)$: if we assume that some small t is the discretization step in time and that the considered functions are constant in time on the interval $]0,t]$, then the factorization [6.130], [6.136], [6.138], [6.139], may be viewed as the continuous spatial factorization of problem $(\mathcal{PP}_0)$ implicitly semi-discretized in time. Then, applying the methods of Chapter 5 yields numerical methods with an LU factorization of the implicit scheme in time.

7

Other Shapes of Domain

In this chapter, we want to show, without completeness, that the success of the invariant embedding method for the factorization of elliptic boundary value problems is not bound to cylindrical domains. The cylindrical case is especially simple as the subdomains are just sub-cylinders and an infinitesimal variation of the domain is achieved by moving the limiting boundary by an infinitesimal amount in the orthogonal direction. In the non-cylindrical case, we need to define, on the one hand, a family of surfaces continuously sweeping over the initial domain and which limit the subdomains, and, on the other hand, a displacement field between these surfaces. In the first step, we study the situation where there is a transformation preserving orthogonality that allows us to reduce the problem to the cylindrical case. Then, the case of subdomains limited by a plane surface but with a non-orthogonal displacement is studied. Further, we turn to the general case where the surfaces of the family have no edge. A particular situation is studied where the domain is star shaped and the subdomains are obtained from the initial domain by a homothety.

7.1. Domain generalization: transformation preserving orthogonal coordinates

Here, we wish to generalize the results obtained in the cylindrical case $\Omega =]0, a[\times \mathcal{O}$ to domains with different shapes. We consider the situation where an orthogonal change of coordinates reduces the problem to a cylinder domain.

7.1.1. *Hypotheses on the domain*

We will use the coordinates $\varphi_i(x_1, x_2, x_3)$, $i = 1, 2, 3$, defined on an open set $\widetilde{\Omega} \subset \mathbb{R}^3$, with $\det(\dfrac{\partial \varphi_i}{\partial x_j})_{i,j=1}^3 \neq 0$. We assume that these coordinates are orthogonal, that is to say:

$$\frac{\partial \varphi_i}{\partial x_1}\frac{\partial \varphi_j}{\partial x_1} + \frac{\partial \varphi_i}{\partial x_2}\frac{\partial \varphi_j}{\partial x_2} + \frac{\partial \varphi_i}{\partial x_3}\frac{\partial \varphi_j}{\partial x_3} = 0, \quad \text{if } i \neq j.$$

We consider the open set Ω in $\mathbb{R}^3$ defined by $\Omega = \{(x_1, x_2, x_3) \mid \varphi_1(x_1, x_2, x_3) \in]0, a[$ and $(\varphi_2(x_1, x_2, x_3), \varphi_3(x_1, x_2, x_3)) \in \mathcal{O}\}$ (assuming that $\overline{\Omega} \subset \widetilde{\Omega}$, where $\overline{\Omega}$ is the closure of Ω). By this change of variables, the domain Ω is transformed into a cylinder. Here are some examples:

EXAMPLE 7.1.– 1) *Cylindrical coordinates.* Let r, θ, x_3 be the cylindrical coordinates given by:

$$\begin{aligned} x_1 &= r \cos\theta \\ x_2 &= r \sin\theta \\ x_3 &= x_3 \end{aligned}$$

and Ω the subset of $\mathbb{R}^3$ whose cylindrical coordinates satisfy $r \in (k_1, k_2)$ $(k_1 > 0)$, $\theta \in (l_1, l_2)$ $(l_2 - l_1 < 2\pi)$ and $x_3 \in (p_1, p_2)$. Then:

$$\begin{cases} \varphi_1(x_1, x_2, x_3) = r(x_1, x_2, x_3) - k_1 \\ \varphi_2(x_1, x_2, x_3) = \theta(x_1, x_2) \\ \varphi_3(x_1, x_2, x_3) = x_3 \end{cases}$$

(see Figure 7.1).

2) *Elliptical coordinates.* Let $c > 0$ be an arbitrary constant. We consider the elliptical coordinates given by:

$$\begin{aligned} x_1 &= c\,\mathrm{ch}\alpha \cos\beta \\ x_2 &= c\,\mathrm{sh}\alpha \sin\beta \\ x_3 &= x_3. \end{aligned}$$

We have the relation:

$$\frac{x_1^2}{c^2\mathrm{ch}^2\alpha} + \frac{x_2^2}{c^2\mathrm{sh}^2\alpha} = 1,$$

then the curve $\alpha = \alpha_0$ with fixed x_3 is an ellipse with half-axis $c\,\mathrm{ch}\alpha_0$ and $c\,\mathrm{sh}\alpha_0$. The distance between foci is $2c$. On the other side:

$$\frac{x_1^2}{c^2\cos^2\beta} - \frac{x_2^2}{c^2\sin^2\beta} = 1,$$

and so the curve $\beta = \beta_0$ with fixed x_3 is a hyperbola with foci $x_1 = \pm c$. Hence, the curves α = constant and β = constant are cylinders whose transverse sections are confocal ellipses and hyperbolas.

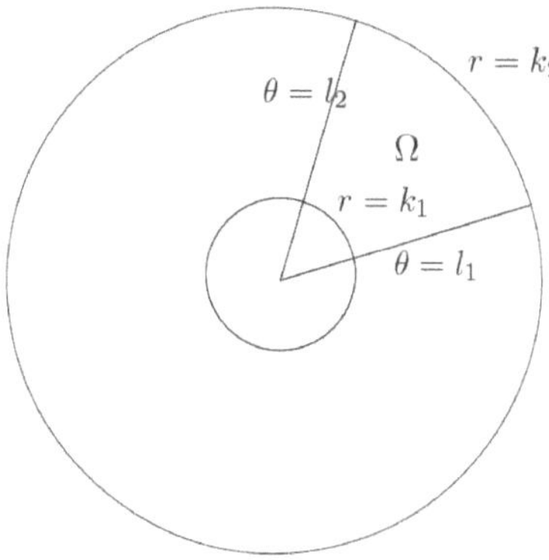

Figure 7.1. *Cylindrical coordinates*

Now, let us consider the open set Ω with elliptical coordinates $\alpha \in (k_1, k_2)$, $\beta \in (l_1, l_2)$ $(l_2 - l_1 < 2\pi)$ and $x_3 \in (p_1, p_2)$. Then:

$$\begin{cases} \varphi_1(x_1, x_2, x_3) = \alpha(x_1, x_2) - k_1 \\ \varphi_2(x_1, x_2, x_3) = \beta(x_1, x_2) \\ \varphi_3(x_1, x_2, x_3) = x_3 \end{cases}$$

(see Figure 7.2).

In a similar way to what was done for the cylinder $]0, a[\times\mathcal{O}$, but applying it now to a domain described in this section, we wish to factorize the following problem:

$$\begin{cases} -\dfrac{\partial^2 u}{\partial x_1^2} - \dfrac{\partial^2 u}{\partial x_2^2} - \dfrac{\partial^2 u}{\partial x_3^2} = f \quad \text{in } \Omega, \\ u|_{\Gamma \setminus (\Gamma_{\varphi_1=0} \cup \Gamma_{\varphi_1=a})} = 0, \\ -\dfrac{\partial u}{\partial \nu}|_{\Gamma_{\varphi_1=0}} = u_0, \quad u|_{\Gamma_{\varphi_1=a}} = u_1, \end{cases} \qquad [7.1]$$

where ν is the outward normal on the boundary, $u_0 \in H_{00}^{1/2}(\Gamma_{\varphi_1=0})'$ and $u_1 \in H_{00}^{1/2}(\Gamma_{\varphi_1=a})$.

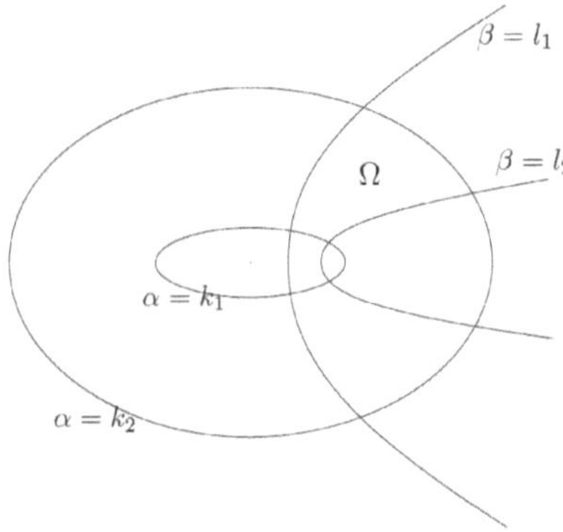

Figure 7.2. *Elliptical coordinates*

7.1.2. *Formal derivation*

Let us rewrite [7.1] in divergence form using the orthogonal coordinates $(\varphi_i)_{i=1,2,3}$:

$$\begin{cases} -\dfrac{\partial}{\partial \varphi_1}((\dfrac{\lambda_2 \lambda_3}{\lambda_1})\dfrac{\partial u}{\partial \varphi_1}) - \dfrac{\partial}{\partial \varphi_2}((\dfrac{\lambda_1 \lambda_3}{\lambda_2})\dfrac{\partial u}{\partial \varphi_2}) - \dfrac{\partial}{\partial \varphi_3}((\dfrac{\lambda_1 \lambda_2}{\lambda_3})\dfrac{\partial u}{\partial \varphi_3}) = \lambda_1 \lambda_2 \lambda_3 f \\ \qquad \text{in }]0, a[\times \mathcal{O}, \\ u|_{\Gamma \setminus (\Gamma_{\varphi_1=0} \cup \Gamma_{\varphi_1=a})} = 0, \; u|_{\Gamma_{\varphi_1=a}} = u_1, \\ -(\dfrac{1}{\lambda_1})\dfrac{\partial u}{\partial \varphi_1}|_{\Gamma_{\varphi_1=0}} = u_0, \end{cases}$$

with $\lambda_i > 0$ and:

$$(\lambda_i)^2 = \left(\frac{\partial x_1}{\partial \varphi_i}\right)^2 + \left(\frac{\partial x_2}{\partial \varphi_i}\right)^2 + \left(\frac{\partial x_3}{\partial \varphi_i}\right)^2$$

(see, for example, Malvern [MAL 77, p. 654]).

We perform the invariant embedding by considering the family of domains Ω_s limited by the mobile boundary $\Omega \bigcap \{\varphi_1 = s\}$. We consider the Neumann-to-Dirichlet operator $Q(s)$ on this mobile boundary.

DEFINITION 7.1.– For each $s \in [0, a[$ and $h \in H_{00}^{1/2}(\mathcal{O})'$, let us define $Q(s)h = \gamma|_s$, where γ is the solution of:

$$\begin{cases} -\dfrac{\partial}{\partial \varphi_1}((\dfrac{\lambda_2\lambda_3}{\lambda_1})\dfrac{\partial \gamma}{\partial \varphi_1}) - \dfrac{\partial}{\partial \varphi_2}((\dfrac{\lambda_1\lambda_3}{\lambda_2})\dfrac{\partial \gamma}{\partial \varphi_2}) - \dfrac{\partial}{\partial \varphi_3}((\dfrac{\lambda_1\lambda_2}{\lambda_3})\dfrac{\partial \gamma}{\partial \varphi_3}) = 0 \\ \qquad\qquad\qquad\qquad\qquad\qquad\qquad\qquad\qquad \text{in }]s, a[\times \mathcal{O}, \\ \gamma|_{\Gamma \backslash (\Gamma_{\varphi_1=0} \cup \Gamma_{\varphi_1=a})} = 0, \ \gamma|_{\Gamma_{\varphi_1=a}} = 0, \\ -(\dfrac{1}{\lambda_1})\dfrac{\partial \gamma}{\partial \varphi_1}|_{\Gamma_{\varphi_1=s}} = h. \end{cases}$$

Likewise, let us define $r(s) = \beta|_{\Gamma_{\varphi=s}}$, where β is the solution of:

$$\begin{cases} -\dfrac{\partial}{\partial \varphi_1}((\dfrac{\lambda_2\lambda_3}{\lambda_1})\dfrac{\partial \beta}{\partial \varphi_1}) - \dfrac{\partial}{\partial \varphi_2}((\dfrac{\lambda_1\lambda_3}{\lambda_2})\dfrac{\partial \beta}{\partial \varphi_2}) - \dfrac{\partial}{\partial \varphi_3}((\dfrac{\lambda_1\lambda_2}{\lambda_3})\dfrac{\partial \beta}{\partial \varphi_3}) = \lambda_1\lambda_2\lambda_3 f \\ \qquad\qquad\qquad\qquad\qquad\qquad\qquad\qquad\qquad \text{in }]s, a[\times \mathcal{O}, \\ \beta|_{\Gamma \backslash (\Gamma_{\varphi_1=0} \cup \Gamma_{\varphi_1=a})} = 0, \ \beta|_{\Gamma_{\varphi_1=a}} = u_1, \\ -(\dfrac{1}{\lambda_1})\dfrac{\partial \beta}{\partial \varphi_1}|_{\Gamma_{\varphi_1=s}} = 0. \end{cases}$$

Finally, set $Q(a) \equiv 0$ and $r(a) = u_1$.

REMARK 7.1.– For each $s \in [0,a]$, $Q(s) : H_{00}^{1/2}(\mathcal{O})' \to H_{00}^{1/2}(\mathcal{O})$ is a linear operator and $r(s) \in H^{1/2}(\mathcal{O})$. Furthermore, the solution u is characterized by:

$$u|_{\varphi_1=s} = -Q(s)(\frac{\lambda_2\lambda_3}{\lambda_1})\frac{\partial u}{\partial \varphi_1}|_{\varphi_1=s} + r(s). \qquad [7.2]$$

Thus, by formally differentiating w.r.t. s, we obtain:

$$\begin{aligned}\frac{\partial u}{\partial \varphi_1}|_{\varphi_1=s} &= -\frac{\mathrm{d}Q}{\mathrm{d}\varphi_1}(s)(\frac{\lambda_2\lambda_3}{\lambda_1})\frac{\partial u}{\partial \varphi_1}|_{\varphi_1=s} \\ &- Q(s)\frac{\partial}{\partial \varphi_1}\left((\frac{\lambda_2\lambda_3}{\lambda_1})\frac{\partial u}{\partial \varphi_1}\right)|_{\varphi_1=s} + \frac{\mathrm{d}r}{\mathrm{d}\varphi_1}(s) \\ &= -\frac{\mathrm{d}Q}{\mathrm{d}\varphi_1}(s)(\frac{\lambda_2\lambda_3}{\lambda_1})\frac{\partial u}{\partial \varphi_1}|_{\varphi_1=s} + Q(s)\frac{\partial}{\partial \varphi_2}\left((\frac{\lambda_1\lambda_3}{\lambda_2})\frac{\partial u}{\partial \varphi_2}\right)|_{\varphi_1=s} \\ &+ Q(s)\frac{\partial}{\partial \varphi_3}\left((\frac{\lambda_1\lambda_2}{\lambda_3})\frac{\partial u}{\partial \varphi_3}\right)|_{\varphi_1=s} \\ &+ Q(s)\lambda_1\lambda_2\lambda_3\, f|_{\varphi_1=s} + \frac{\mathrm{d}r}{\mathrm{d}\varphi_1}(s). \end{aligned} \qquad [7.3]$$

Then:

$$\begin{aligned}\frac{\partial u}{\partial \varphi_1}|_{\varphi_1=s} &= -\frac{\mathrm{d}Q}{\mathrm{d}\varphi_1}(s)(\frac{\lambda_2\lambda_3}{\lambda_1})\frac{\partial u}{\partial \varphi_1}|_{\varphi_1=s} \\ &- Q(s)\frac{\partial}{\partial \varphi_2}\left((\frac{\lambda_1\lambda_3}{\lambda_2})\frac{\partial}{\partial \varphi_2}Q(s)(\frac{\lambda_2\lambda_3}{\lambda_1})\frac{\partial u}{\partial \varphi_1}|_{\varphi_1=s}\right) \\ &- Q(s)\frac{\partial}{\partial \varphi_3}\left((\frac{\lambda_1\lambda_2}{\lambda_3})\frac{\partial}{\partial \varphi_3}Q(s)(\frac{\lambda_2\lambda_3}{\lambda_1})\frac{\partial u}{\partial \varphi_1}|_{\varphi_1=s}\right) \\ &+ Q(s)\frac{\partial}{\partial \varphi_2}\left((\frac{\lambda_1\lambda_3}{\lambda_2})\frac{\partial}{\partial \varphi_2}r\right) + Q(s)\frac{\partial}{\partial \varphi_3}(\frac{\lambda_1\lambda_2}{\lambda_3})\frac{\partial}{\partial \varphi_3}r \\ &\qquad + Q(s)\lambda_1\lambda_2\lambda_3\, f|_{\varphi_1=s} + \frac{\mathrm{d}r}{\mathrm{d}\varphi}(s). \end{aligned} \qquad [7.4]$$

Hence, in a similar way to the cylindrical case, since $\dfrac{\partial u}{\partial \varphi_1}$ is arbitrary, we obtain the Riccati equation for Q:

$$\begin{cases} \dfrac{\mathrm{d}Q}{\mathrm{d}\varphi_1}(\dfrac{\lambda_2\lambda_3}{\lambda_1})I + Q(\varphi_1) \circ \left[\dfrac{\partial}{\partial \varphi_2}\left((\dfrac{\lambda_1\lambda_3}{\lambda_2})\dfrac{\partial}{\partial \varphi_2}\right) + \dfrac{\partial}{\partial \varphi_3}\left((\dfrac{\lambda_1\lambda_2}{\lambda_3})\dfrac{\partial}{\partial \varphi_3}\right)\right], \\ \qquad\qquad\qquad \circ Q(\varphi_1)(\dfrac{\lambda_2\lambda_3}{\lambda_1})I + I = 0 \\ Q(a) = 0. \end{cases}$$

and the equation for the residual r:

$$\begin{cases} \dfrac{\mathrm{d}r}{\mathrm{d}\varphi_1} + Q(\varphi_1)\left(\dfrac{\partial}{\partial \varphi_2}(\dfrac{\lambda_1\lambda_3}{\lambda_2})\dfrac{\partial}{\partial \varphi_2}r + \dfrac{\partial}{\partial \varphi_3}(\dfrac{\lambda_1\lambda_2}{\lambda_3})\dfrac{\partial}{\partial \varphi_3}r\right) \\ \qquad\qquad +Q(\varphi_1)\lambda_1\lambda_2\lambda_3\, f|_{\varphi_1=s} = 0, \\ r(a) = u_1. \end{cases}$$

In conclusion, we obtained the following result:

THEOREM 7.1.– The solution of problem [7.1] is determined by the decoupled system:

$$\begin{cases} \dfrac{\mathrm{d}Q}{\mathrm{d}\varphi_1}(\dfrac{\lambda_2\lambda_3}{\lambda_1})I + Q(\varphi_1) \circ \left[\dfrac{\partial}{\partial \varphi_2}\left((\dfrac{\lambda_1\lambda_3}{\lambda_2})\dfrac{\partial}{\partial \varphi_2}\right) + \dfrac{\partial}{\partial \varphi_3}\left((\dfrac{\lambda_1\lambda_2}{\lambda_3})\dfrac{\partial}{\partial \varphi_3}\right)\right] \\ \qquad\qquad\qquad \circ Q(\varphi_1)(\dfrac{\lambda_2\lambda_3}{\lambda_1})I + I = 0, \\ Q(a) = 0, \\ \dfrac{\mathrm{d}r}{\mathrm{d}\varphi} + Q(\varphi_1)\left(\dfrac{\partial}{\partial \varphi_2}(\dfrac{\lambda_1\lambda_3}{\lambda_2})\dfrac{\partial}{\partial \varphi_2}r + \dfrac{\partial}{\partial \varphi_3}(\dfrac{\lambda_1\lambda_2}{\lambda_3})\dfrac{\partial}{\partial \varphi_3}r\right) \\ \qquad\qquad +Q(\varphi_1)\lambda_1\lambda_2\lambda_3\, f|_{\varphi_1=s} = 0, \\ r(a) = u_1, \\ Q(\varphi_1)\dfrac{\lambda_2\lambda_3}{\lambda_1}\dfrac{\mathrm{d}u}{\mathrm{d}\varphi_1} + u = r, \\ u(0) = Q(0)\lambda_2\lambda_3 u_0 + r(0). \end{cases}$$

7.2. Quasi-cylindrical domains with normal velocity fields

Let us consider the problem associated with the Laplacian in an open set Ω, which is quasi-cylindrical with respect to the x-axis.

We call quasi-cylindrical a domain Ω defined by its sections $\Omega = \bigcup_{x=0}^{a}(\{x\}, \mathcal{O}_x)$ (see Figure 7.3). The sections $\mathcal{O}_x$ are bounded sets in $\mathbb{R}^{n-1}$ with a regular boundary $\partial\mathcal{O}_x$ depending regularly on x, without topology change. Let $\Sigma = \bigcup_{x=0}^{a}(\{x\}, \partial\mathcal{O}_x)$ be the lateral boundary of Ω. Let n denote the outward normal to Ω on Σ and let ν be the outward normal to $\mathcal{O}_x$ in the section plane at x in $\mathbb{R}^{n-1}$ at a point of $\partial\mathcal{O}_x$. The domain Ω is bounded by the faces $\Gamma_0 = \{0\} \times \mathcal{O}_0$ and $\Gamma_a = \{a\} \times \mathcal{O}_a$ and the lateral boundary Σ.

We consider the following problem:

$$(\mathcal{P}_0)\begin{cases} -\Delta u = f \quad \text{in } \Omega, \quad f \in L^2(\Omega), \\ u|_\Sigma = 0, \quad -\dfrac{\partial u}{\partial n}|_{\Gamma_0} = u_0 \in (H^{1/2}_{0,0}(\mathcal{O}_0))', \\ u|_{\Gamma_a} = u_1 \in H^{1/2}_{0,0}(\mathcal{O}_a). \end{cases} \qquad [7.5]$$

Here, we present a factorization linked to a natural way of moving the mobile section in the process of invariant embedding: each point is moving normally to the section. When dealing with example 7.4 in section 7.3, we will come back to this problem but with a different way of moving the section that provides a different factorization of the same problem.

Let $\mathcal{O}$ be a fixed neighborhood of $\mathcal{O}_x : \mathcal{O}_x \subset \mathcal{O}, \forall x \in [0, a]$. We define the cylinder $\tilde{\Omega} =]0, a[\times\mathcal{O}$, $\Omega \subset \tilde{\Omega}$. The function u is extended by 0 in $\tilde{\Omega}$.

As before, we embed this problem in the family:

$$(\mathcal{P}_{s,h})\begin{cases} -\Delta u_s = f \quad \text{in } \Omega_s, \\ u_s|_{\Sigma_s} = 0, \quad -\dfrac{\partial u_s}{\partial n}|_{\Gamma_0} = u_0, \quad u_s|_{\Gamma_s} = h \in H^{1/2}_{0,0}(\mathcal{O}_s), \end{cases}$$

where $\Omega_s = \Omega \bigcap \{x < s\}$, $\Sigma_s = \Sigma \bigcap \{x < s\}$ and h is an arbitrary data. By linearity of $(\mathcal{P}_{s,h})$, we have:

$$\frac{\partial u_s}{\partial x}|_{\Gamma_s} = P(s)h + w(s), \qquad [7.6]$$

where the Dirichlet-to-Neumann operator $P(s)$ and the residual $w(s)$ are defined, as in section 1.2. Let us consider now the restriction of $u_s(h)$ to Ω_x for $x \leq s$. It is a solution of problem $(\mathcal{P}_{x,u_s(x;h)})$. By the previous reasoning, we obtain the relation:

$$\frac{\mathrm{d}u_s}{\mathrm{d}x}(x;h) = P(x)u_s(x;h) + w(x). \qquad [7.7]$$

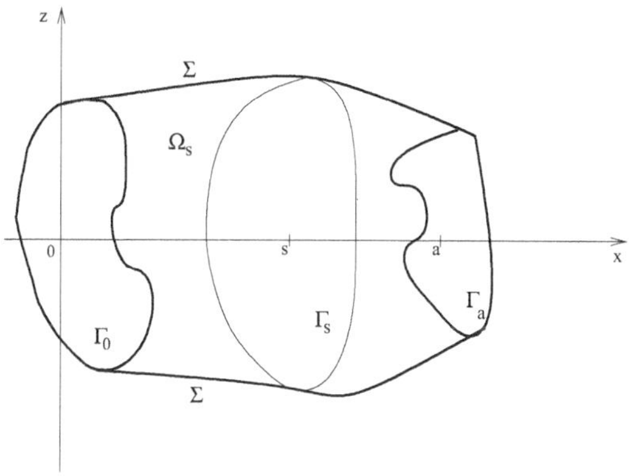

Figure 7.3. *Quasi-cylindrical domain*

At this level, we cannot directly compute the derivative with respect to x because h is defined on $\mathcal{O}_x$, which depends on x. We use the extension of u by 0 to $\tilde{\Omega}$ and in the same way for w and P.

Let $k \in H_0^1(\mathcal{O})$ be an arbitrary fixed test function, and take h in $H_0^1(\mathcal{O}_s)$. Integrating over $\tilde{\Omega}$:

$$\int_{\mathcal{O}} \frac{\mathrm{d}u_s}{\mathrm{d}x}(x;h)k \, \mathrm{d}y = \int_{\mathcal{O}} (P(x)u_s(x;h) + w(x))k \, \mathrm{d}y. \qquad [7.8]$$

For the left-hand side:

$$\begin{aligned}\frac{\mathrm{d}}{\mathrm{d}x}\int_{\mathcal{O}} \frac{\mathrm{d}u_s}{\mathrm{d}x}k \, \mathrm{d}y &= \int_{\mathcal{O}} \frac{\mathrm{d}^2 u_s}{\mathrm{d}x^2}k \, \mathrm{d}y \\ &= -\int_{\mathcal{O}} \Delta_y u_s k \, \mathrm{d}y - \int_{\mathcal{O}} fk \, \mathrm{d}y.\end{aligned}$$

For the right-hand side:

$$\frac{\mathrm{d}}{\mathrm{d}x}\int_{\mathcal{O}}(P(x)u_s(x)+w(x))k\ \mathrm{d}y = \int_{\mathcal{O}}\frac{\mathrm{d}}{\mathrm{d}x}(P(x)u_s(x)+w(x))\,k\,\mathrm{d}y$$
$$= \int_{\mathcal{O}}(\frac{\mathrm{d}P}{\mathrm{d}x}u_s(x)+P(x)\frac{\mathrm{d}u_s}{\mathrm{d}x}(x)+\frac{\mathrm{d}w}{\mathrm{d}x})k\ \mathrm{d}y.$$

By substitution using [7.7]:

$$\frac{\mathrm{d}}{\mathrm{d}x}\int_{\mathcal{O}}(P(x)u_s(x)+w(x))k\ \mathrm{d}y = \int_{\mathcal{O}}(\frac{\mathrm{d}P}{\mathrm{d}x}+P(x)^2)u_s(x)k\ \mathrm{d}y$$
$$+\int_{\mathcal{O}}(\frac{\mathrm{d}w}{\mathrm{d}x}+Pw)k\ \mathrm{d}y.$$

Equating both sides:

$$\int_{\mathcal{O}}(\frac{\mathrm{d}P}{\mathrm{d}x}+P(x)^2+\Delta_y)u_s(x)k\ \mathrm{d}y+\int_{\mathcal{O}}(\frac{\mathrm{d}w}{\mathrm{d}x}+Pw+f)k\ \mathrm{d}y=0,$$

which is valid for $0 \leq x \leq s$. Now, choosing $x = s$, let us come back to the restriction to $\mathcal{O}_s$ (since integrands are zero in the complement) and take k in $H_0^1(\mathcal{O}_s)$ to be the test function. We obtain:

$$(\frac{\mathrm{d}P}{\mathrm{d}x}h,k)_{\mathcal{O}_s}+(Ph,Pk)_{\mathcal{O}_s}=(\nabla_y h,\nabla_y k)_{\mathcal{O}_s},\quad \forall h,k\in H_0^1(\mathcal{O}_s),\quad [7.9]$$

(where $(.,.)_{\mathcal{O}_s}$ is the scalar product in $L^2(\mathcal{O}_s)$), which is the variational formulation for the Riccati equation for P in the non-cylindrical open set Ω. In the same way for the residual w:

$$\int_{\mathcal{O}_s}(\frac{\mathrm{d}w}{\mathrm{d}x}+Pw+f)k\ \mathrm{d}y=0,\quad \forall k\in H_0^1(\mathcal{O}_s).\quad [7.10]$$

Now, we have to determine the boundary conditions satisfied by P and w on Σ. The same computation as before, but using test functions that do not necessarily vanish on the boundary, would lead us to study the traces of w or P on the boundary $\partial\mathcal{O}_x$, which would need additional regularities. We resume the previous computation by integrating in x within Ω_s with a test function v chosen in $H^1(\Omega_s)$ that does not necessarily vanish on Σ_s. Let M be a point of Σ. Let $\vec{\tau}$ be the tangent vector to Σ in the plane defined by the parallel to $\vec{Ox}$ through M and the outward normal ν to $\mathcal{O}_x$. Let α be the angle $(\vec{Ox}, \vec{\tau})$. The normal speed of deformation of $\partial\mathcal{O}_x$ is $\tan\alpha$. Then, we can take the derivative of both sides of [7.8] with respect to x using the formulas from [SOK 92]. We obtain:

$$\begin{aligned}\int_{\mathcal{O}_s} \frac{\partial u}{\partial x} v \, \mathrm{d}y + \int_{\mathcal{O}_0} u_0 v \, \mathrm{d}y &= \int_0^s \frac{\mathrm{d}}{\mathrm{d}x} \int_{\mathcal{O}_x} \frac{\partial u}{\partial x} v \, \mathrm{d}x \, \mathrm{d}y \\ &= \int_{\Omega_s} \frac{\partial^2 u}{\partial x^2} v \, \mathrm{d}x \, \mathrm{d}y + \int_{\Omega_s} \frac{\partial u}{\partial x} \frac{\partial v}{\partial x} \, \mathrm{d}x \, \mathrm{d}y \\ &+ \int_0^s \int_{\partial\mathcal{O}_x} \frac{\partial u}{\partial x} v \tan\alpha \, \mathrm{d}l \, \mathrm{d}x.\end{aligned}$$

The last integral is meaningful because given $\mathrm{d}l$, the arc length along $\partial\mathcal{O}_x$, the surface element of Σ satisfies $\cos\alpha \, \mathrm{d}\Sigma = \mathrm{d}l \, \mathrm{d}x$. Hence, we have:

$$\int_0^s \int_{\partial\mathcal{O}_x} \frac{\partial u}{\partial x} v \tan\alpha \, \mathrm{d}l \, \mathrm{d}x = \int_{\Sigma_s} \frac{\partial u}{\partial x} v \sin\alpha \, \mathrm{d}\Sigma. \qquad [7.11]$$

On the other hand, starting from the equation of problem $(\mathcal{P}_{s,h})$, Green's formula gives:

$$\begin{aligned}\int_{\mathcal{O}_s} \frac{\partial u}{\partial x} v \, \mathrm{d}y + \int_{\mathcal{O}_0} u_0 v \, \mathrm{d}y = \int_{\Omega_s} \nabla u \nabla v \, \mathrm{d}x \, \mathrm{d}y - \int_{\Omega_s} f v \, \mathrm{d}x \, \mathrm{d}y \\ - \int_{\Sigma_s} \frac{\partial u}{\partial n} v \, \mathrm{d}\Sigma, \qquad [7.12]\end{aligned}$$

and so by equating these two expressions and after simplification:

$$\begin{aligned}\int_{\Omega_s} \frac{\partial^2 u}{\partial x^2} v \, \mathrm{d}x \, \mathrm{d}y + \int_{\Sigma_s} \frac{\partial u}{\partial x} v \sin\alpha \, \mathrm{d}\Sigma = \int_{\Omega_s} \nabla_y u \nabla_y v \, \mathrm{d}x \, \mathrm{d}y \\ - \int_{\Omega_s} f v \, \mathrm{d}x \, \mathrm{d}y - \int_{\Sigma_s} \frac{\partial u}{\partial n} v \, \mathrm{d}\Sigma. \qquad [7.13]\end{aligned}$$

Therefore, by [7.7], using Green's formula on each section, we obtain:

$$\begin{aligned}&\int_{\Omega_s} \frac{\mathrm{d}}{\mathrm{d}x}(Pu + w)v \, \mathrm{d}x \, \mathrm{d}y + \int_{\Sigma_s} (Pu + w)v \sin\alpha \, \mathrm{d}\Sigma \\ &= \int_{\Omega_s} \nabla_y u \nabla_y v \, \mathrm{d}x \, \mathrm{d}y - \int_{\Omega_s} fv \, \mathrm{d}x \, \mathrm{d}y - \int_{\Sigma_s} \frac{\partial u}{\partial n} v \, \mathrm{d}\Sigma \\ &= - \int_{\Omega_s} \Delta_y uv \, \mathrm{d}x \, \mathrm{d}y + \int_0^s \int_{\partial \mathcal{O}_x} \frac{\partial u}{\partial \nu} v \, \mathrm{d}l \, \mathrm{d}x - \int_{\Omega_s} fv \, \mathrm{d}x \, \mathrm{d}y \\ &\quad - \int_{\Sigma_s} \frac{\partial u}{\partial n} v \, \mathrm{d}\Sigma. \end{aligned} \qquad [7.14]$$

By a computation similar to the previous one:

$$\begin{aligned}\int_{\Omega_s} \frac{\mathrm{d}}{\mathrm{d}x}(Pu + w)v \, \mathrm{d}x \, \mathrm{d}y &= \int_{\Omega_s} (\frac{\mathrm{d}P}{\mathrm{d}x} + P^2)uv \, \mathrm{d}x \, \mathrm{d}y \\ &+ \int_{\Omega_s} (\frac{\mathrm{d}w}{\mathrm{d}x} + Pw)v \, \mathrm{d}x \, \mathrm{d}y. \end{aligned} \qquad [7.15]$$

Moreover, by Green's formula applied to the section, [7.9] reads:

$$(\frac{\mathrm{d}P}{\mathrm{d}x} h, k)_{\mathcal{O}_x} + (Ph, Pk)_{\mathcal{O}_x} = -(\Delta_y h, k)_{\mathcal{O}_x} \qquad [7.16]$$

If we assume that $\Delta_y h \in L^2(\mathcal{O}_x)$, this formula can be extended to any $k \in L^2(\mathcal{O}_x)$ by denseness. Let us choose $h = u(x), k = v(x)$ and integrate from 0 to s. By transferring this result and [7.10] into [7.14], we obtain:

$$\int_{\Sigma_s} (Pu + w)v \sin\alpha \, \mathrm{d}\Sigma = \int_{\Sigma_s} \frac{\partial u}{\partial \nu} v \cos\alpha \, \mathrm{d}\Sigma - \int_{\Sigma_s} \frac{\partial u}{\partial n} v \, \mathrm{d}\Sigma. \qquad [7.17]$$

Assume that u and v are functions $C^1(\Omega_s)$. Using the fact that $u|_\Sigma = 0$ and the regularity of Σ:

$$\frac{\partial u}{\partial \nu}|_\Sigma = \cos\alpha \frac{\partial u}{\partial n}|_\Sigma,$$

we have that:

$$\begin{aligned}&\int_{\Sigma_s} \frac{\partial u}{\partial \nu} v \cos\alpha \, \mathrm{d}\Sigma - \int_{\Sigma_s} \frac{\partial u}{\partial n} v \, \mathrm{d}\Sigma \\ &\qquad = - \int_{\Sigma_s} \frac{\partial u}{\partial n} v \sin^2\alpha \, \mathrm{d}\Sigma = - \int_{\Sigma_s} \tan\alpha \frac{\partial u}{\partial \nu} v \sin\alpha \, \mathrm{d}\Sigma.\end{aligned}$$

and we conclude that:

$$\int_{\Sigma_s} (Pu + w)v \sin\alpha \, \mathrm{d}\Sigma = -\int_{\Sigma_s} \tan\alpha \frac{\partial u}{\partial \nu} v \sin\alpha \, \mathrm{d}\Sigma, \qquad [7.18]$$

This formula can be extended by denseness to any function u in $L^2(0, a; H_0^1(\mathcal{O}_x))$, such that $\Delta_y u \in L^2(\Omega)$, and any function $v \in L^2(0, a; H_0^1(\mathcal{O}_x))$. From this identity, we conclude that, in particular, taking $u = 0$:

$$\int_{\Sigma_s} wv \sin\alpha \, \mathrm{d}\Sigma = 0, \quad \forall v \in L^2(0, a; H^{1/2}(\partial\mathcal{O}_x)), \forall s. \qquad [7.19]$$

Therefore, almost everywhere on Σ, $w = 0$ whenever $\alpha \neq 0$. In a complementary way, the boundary condition of P reads:

$$\int_{\Sigma_s} (Pu + \tan\alpha \frac{\partial u}{\partial \nu}) v \sin\alpha \, \mathrm{d}\Sigma = 0. \qquad [7.20]$$

Therefore, on almost every section $\mathcal{O}_x$, $\gamma_x \circ P = -\tan\alpha \frac{\partial}{\partial \nu}$, whenever $\alpha \neq 0$, where γ_x is the trace operator on the boundary $\partial\mathcal{O}_x$.

7.3. Sweeping the domain by surfaces of arbitrary shape

In this section, we consider the factorization of a boundary value problem for which the domain and the family of surfaces of the invariant embedding have no particular shape. We are looking for a formulation of this factorization as intrinsic as possible, that is, independent of a coordinate system linked to the geometry. To avoid an edge effect when the displacement field is not tangent to the boundary such as that exhibited in section 7.2, we assume that the surfaces defining the invariant embedding have no edge. Without loss of generality, we will take $a = 1$ in the section.

Let $\tilde{\Omega}_0$ and $\tilde{\Omega}_1$ be two open sets in $\mathbb{R}^3$ (see Figure 7.4) with boundaries, respectively, given by Γ_0 and Γ_1 of class $\mathcal{C}^1$ with $\overline{\tilde{\Omega}_0} \subset \tilde{\Omega}_1$. Let $\{\Gamma_s\}_{s=0}^{s=1}$ be a family of surfaces of class $\mathcal{C}^1$ with a $\mathcal{C}^1$-dependency on s and that are the respective boundaries of the open sets of the family $\tilde{\Omega}_s$ with:

$$\overline{\tilde{\Omega}_s} \subset \tilde{\Omega}_{s'}, \text{ for } 0 \leq s < s' \leq 1.$$

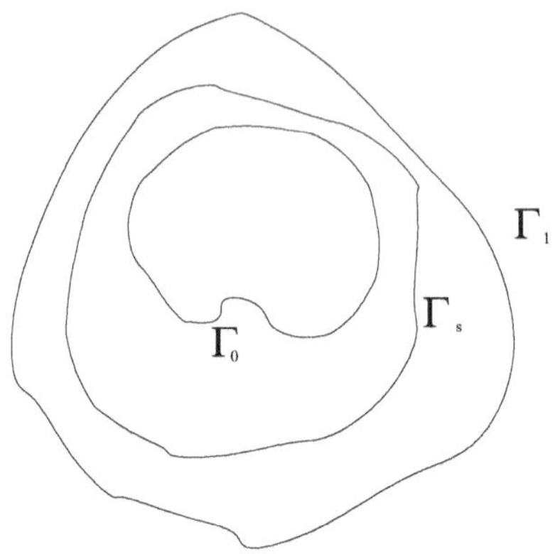

Figure 7.4. *Interlocking of domains for the invariant embedding*

The domain Ω is chosen as:

$$\Omega = \tilde{\Omega}_1 \setminus \overline{\tilde{\Omega}_0},$$

and we set:

$$\Omega_s = \tilde{\Omega}_s \setminus \overline{\tilde{\Omega}_0}.$$

Let $\Phi_s^{s'}$ be the C^1 canonical diffeomorphism from Γ_s onto $\Gamma_{s'}$. We take, for example, Γ_0 as a reference set and let t be a coordinate system on Γ_0 defined by local maps. We thus obtain a local coordinate system (s, t). Let $\alpha(x)$ and $V_{\Gamma_s}(x)$ denote the projections of $V(x) = \dfrac{\mathrm{d}}{\mathrm{d}s}\Phi_0^s(x)$ on the outward normal to Ω_s and the linear tangent space to Γ_s at $x \in \Gamma_s$, respectively.

Let $\nabla_{\Gamma_s} u$ denote the projection of ∇u on the tangent space to Γ_s at $x \in \Gamma_s$. It is known that this tangential gradient ∇_{Γ_s} is independent of the system of local maps of the manifold Γ_s. From the previous definitions of α and V_{Γ_s}, we have:

$$\frac{\partial u}{\partial s} = \alpha \frac{\partial u}{\partial n} + V_{\Gamma_s} \cdot \nabla_{\Gamma_s} u. \qquad [7.21]$$

Let us consider the following form of problem $(\mathcal{P}_0)$:

$$(\mathcal{P}_0) \begin{cases} -\Delta u = f \quad \text{in } \Omega, \\ \dfrac{\partial u}{\partial n}|_{\Gamma_0} = u_0 \in H^{-1/2}(\Gamma_0), \quad u|_{\Gamma_1} = u_1 \in H^{1/2}(\Gamma_1), \end{cases} \qquad [7.22]$$

where n is the outward normal to Ω. Here, it is important to note that the outward normal to Ω on Γ_0 is opposite to the outward normal to Ω_0, which appears in [7.21]. The invariant embedding uses the family:

$$(\mathcal{P}_{s,h})\begin{cases} -\Delta u = f \quad \text{in}\, \Omega_s, \\ \dfrac{\partial u}{\partial n}|_{\Gamma_0} = u_0, \quad u|_{\Gamma_s} = h \in H^{1/2}(\Gamma_s). \end{cases} \qquad [7.23]$$

REMARK 7.2.– Unlike the model case of section 1.2, the Dirichlet-to-Neumann operator is not one-to-one and its kernel is composed of constant functions on Γ_s. Indeed, if we add a constant C to the solution of [7.23], that is to say if we consider the solution of this problem with the Dirichlet data $h + C$, then the corresponding value of the Neumann condition $\dfrac{\partial u}{\partial n}|_{\Gamma_s}$ is unchanged. This is due to the fact that here Γ_s has no edge and so the Dirichlet condition on Σ that ensured the one-to-one property is no longer present. Also, the operator P is not surjective, because in order for the corresponding Neumann problem, that is, with the conditions $\dfrac{\partial u}{\partial n}|_{\Gamma_0} = 0$ and $\dfrac{\partial u}{\partial n}|_{\Gamma_s} = P(s)h$, to be well-posed, we require that $\int_{\Gamma_s} P(s)h\, d\Gamma_s = 0$. It can be shown that $P(s)$ is an isomorphism from $\{u \in H^{1/2}(\Gamma_s) | \int_{\Gamma_s} u\, d\Gamma_s = 0\}$ onto $\{u \in H^{-1/2}(\Gamma_s) | \int_{\Gamma_s} u\, d\Gamma_s = 0\}$. The situation is similar to that of section 3.2.1 for a Neumann condition on the lateral surface of the cylinder.

We proceed to the formal computation following the method presented in section 3.1. Let $v \in H^1(\Omega_{s0})$ be a test function. Applying Green's formula to [7.23]:

$$\int_{\Omega_{s0}} \nabla u \cdot \nabla v \, dx = \int_{\Omega_{s0}} f v \, dx + \int_{\Gamma_{s0}} \frac{\partial u}{\partial n} v \, d\Gamma_{s0} + \int_{\Gamma_0} u_0 v \, d\Gamma_0. \quad [7.24]$$

Let us consider the left-hand side, and decompose the gradient into its components normal and tangential to Γ_s. From [7.21], we have:

$$\begin{aligned} \int_{\Omega_{s0}} \nabla u \cdot \nabla v \, dx &= \int_0^{s0} \int_{\Gamma_s} \left(\frac{\partial u}{\partial n}\frac{\partial v}{\partial n} + \nabla_{\Gamma_s} u \cdot \nabla_{\Gamma_s} v \right) \alpha \, ds \, d\Gamma_s \\ = \int_0^{s0} \int_{\Gamma_s} &\left(\frac{1}{\alpha^2} \left(\frac{\partial u}{\partial s}\frac{\partial v}{\partial s} - (\frac{\partial u}{\partial s} V_{\Gamma_s} \cdot \nabla_{\Gamma_s} v + \frac{\partial v}{\partial s} V_{\Gamma_s} \cdot \nabla_{\Gamma_s} u) \right) \right. \\ &\left. + \frac{1}{\alpha^2} V_{\Gamma_s} \cdot \nabla_{\Gamma_s} u V_{\Gamma_s} \cdot \nabla_{\Gamma_s} v + \nabla_{\Gamma_s} u \cdot \nabla_{\Gamma_s} v \right) \alpha \, ds \, d\Gamma_s. \quad [7.25] \end{aligned}$$

Now we choose v independent of s:

$$\int_{\Omega_{s_0}} \nabla u \cdot \nabla v \, \mathrm{d}x = \int_0^{s_0} \int_{\Gamma_s} \left(-\frac{1}{\alpha^2} \frac{\partial u}{\partial s} V_{\Gamma_s} \cdot \nabla_{\Gamma_s} v \right.$$
$$\left. + (I + \frac{1}{\alpha^2} V_{\Gamma_s} \otimes V_{\Gamma_s}) \nabla_{\Gamma_s} u \cdot \nabla_{\Gamma_s} v \right) \alpha \, \mathrm{d}s \, \mathrm{d}\Gamma_s. \quad [7.26]$$

Differentiating this equality with respect to s_0 (and returning to the notation $s_0 = s$), we have that:

$$\frac{\mathrm{d}}{\mathrm{d}s} \int_{\Omega_s} \nabla u \cdot \nabla v \, \mathrm{d}x = \int_{\Gamma_s} \left(-\frac{1}{\alpha^2} \frac{\partial u}{\partial s} V_{\Gamma_s} \cdot \nabla_{\Gamma_s} v \right.$$
$$\left. + (I + \frac{1}{\alpha^2} V_{\Gamma_s} \otimes V_{\Gamma_s}) \nabla_{\Gamma_s} u \cdot \nabla_{\Gamma_s} v \right) \alpha \, \mathrm{d}\Gamma_s \quad [7.27]$$
$$= \int_{\Gamma_s} \left(-\frac{1}{\alpha} \frac{\partial u}{\partial n} V_{\Gamma_s} \cdot \nabla_{\Gamma_s} v + \nabla_{\Gamma_s} u \cdot \nabla_{\Gamma_s} v \right) \alpha \, \mathrm{d}\Gamma_s,$$

as a result of [7.21]. Let $(.,.)_{\Gamma_x}$ denote the scalar product in $L^2(\Gamma_x)$ or the duality $H^{1/2}(\Gamma_x) \times H^{-1/2}(\Gamma_x)$. By the usual reasoning, we link the normal derivative on Γ_s to the data h on the same surface by:

$$(\frac{\partial u}{\partial n}, v)_{\Gamma_s} = (P(s)h + w(s), v)_{\Gamma_s}. \quad [7.28]$$

Therefore, for the solution of $(\mathcal{P}_{s,h})$, we have:

$$(\frac{\partial u}{\partial n}|_{\Gamma_s}, v)_{\Gamma_s} = (P(s)u|_{\Gamma_s} + w(s), v)_{\Gamma_s}. \quad [7.29]$$

Transferring to [7.27]:

$$\frac{\mathrm{d}}{\mathrm{d}s} \int_{\Omega_s} \nabla u \cdot \nabla v \, \mathrm{d}x = \int_{\Gamma_s} \left(-\frac{1}{\alpha} (P(s)u + w(s)) V_{\Gamma_s} \cdot \nabla_{\Gamma_s} v \right.$$
$$\left. + \nabla_{\Gamma_s} u \cdot \nabla_{\Gamma_s} v \right) \alpha \, \mathrm{d}\Gamma_s. \quad [7.30]$$

The same applies to differentiating the right-hand side of [7.24]; for the first term, we have:

$$\frac{\mathrm{d}}{\mathrm{d}s}\int_{\Omega_s} fv\,\mathrm{d}x = \int_{\Gamma_s} fv\,\alpha\,\mathrm{d}\Gamma_s. \qquad [7.31]$$

The last term depends only on s. In order to derive the integral on Γ_s, we use the formula by Delfour and Zolésio [DEL 01, p. 355]:

$$\begin{aligned}&\frac{\mathrm{d}}{\mathrm{d}s}\left(\int_{\Gamma_s(V)} \psi(s,x)\,\mathrm{d}\Gamma_s\right)|_{s=0} = \\ &= \int_{\Gamma_0} (\psi'(0,x) + \nabla\psi(0,x)\cdot V(0) \\ &\quad + \psi(0,x)(\mathrm{div}V(0) - DV(0)n\cdot n))\,\mathrm{d}\Gamma_0,\end{aligned} \qquad [7.32]$$

where DV denotes the Jacobian of V.

In formula [7.32], the notation $\psi(s)$ refers to a direct and explicit dependency of ψ with respect to s and ψ' is the corresponding derivative. This is not useful in our computation. On the other hand, the term $\nabla\psi\cdot V$ represents the derivative of ψ with respect to s with t fixed and is linked to a variation of the geometry of the domain. For $\psi = v$, this derivative is zero by construction:

$$\nabla v\cdot V(0) = 0.$$

Let us set:

$$\beta = \mathrm{div}V(0) - DV(0)n\cdot n.$$

In formula [7.32], the term in β is linked to the variation in s of the element of integration $\mathrm{d}\Gamma_s$.

We obtain:

$$
\begin{aligned}
&\frac{\mathrm{d}}{\mathrm{d}s}\left(\int_{\Gamma_s}(P(s)u(s)+w(s))v\,\mathrm{d}\Gamma_s\right)=\\
&=\int_{\Gamma_s}\frac{\partial}{\partial s}(P(s)u(s)+w(s))v\,\mathrm{d}\Gamma_s+\int_{\Gamma_s}\beta(P(s)u(s)+w(s))v\,\mathrm{d}\Gamma_s\\
&=\int_{\Gamma_s}(\frac{\mathrm{d}P}{\mathrm{d}s}u+P(\alpha\frac{\partial u}{\partial n}+V_{\Gamma_s}\cdot\nabla_{\Gamma_s}u)\\
&\quad+\frac{\partial w}{\partial s}+\beta(P(s)u(s)+w(s)))v\,\mathrm{d}\Gamma_s\\
&=\int_{\Gamma_s}(\frac{\mathrm{d}P}{\mathrm{d}s}+P\alpha P+PV_{\Gamma_s}\cdot\nabla_{\Gamma_s}+\beta P)uv\,\mathrm{d}\Gamma_s\\
&\quad+\int_{\Gamma_s}\left(\frac{\partial w}{\partial s}+P\alpha w+\beta w\right)v\,\mathrm{d}\Gamma_s.
\end{aligned}
\tag{7.33}
$$

By the argument of "arbitrary trajectory" of invariant embedding as in section 1.2 and using the fact that P is self-adjoint, which can be proven as in proposition 1.2, we deduce the Riccati equation satisfied by P and the equation for the residual w:

$$
\begin{aligned}
&(\frac{\mathrm{d}P}{\mathrm{d}s}u,v)_{\Gamma_s}+(\alpha Pu,Pv)_{\Gamma_s}+(PV_{\Gamma_s}\cdot\nabla_{\Gamma_s}u,v)_{\Gamma_s}\\
&+(u,PV_{\Gamma_s}\cdot\nabla_{\Gamma_s}v)_{\Gamma_s}+(\beta Pu,v)_{\Gamma_s}=\int_{\Gamma_s}\alpha\nabla_{\Gamma_s}u\cdot\nabla_{\Gamma_s}v\,\mathrm{d}\Gamma_s,\\
&P(0)=0,
\end{aligned}
\tag{7.34}
$$

$$
\begin{aligned}
&(\frac{\partial w}{\partial s},v)_{\Gamma_s}+(P\alpha w,v)_{\Gamma_s}+(w,V_{\Gamma_s}\cdot\nabla_{\Gamma_s}v)_{\Gamma_s}+(\beta w,v)_{\Gamma_s}+(\alpha f,v)_{\Gamma_s}=0,\\
&w(0)=-u_0.
\end{aligned}
\tag{7.35}
$$

Thanks to [7.21], relation [7.29] gives for u:

$$
(\frac{\partial u}{\partial s},v)_{\Gamma_s}=(\alpha Pu,v)_{\Gamma_s}+(V_{\Gamma_s}\cdot\nabla_{\Gamma_s}u,v)_{\Gamma_s}+(\alpha w,v)_{\Gamma_s},\quad u(1)=u_1. \tag{7.36}
$$

REMARK 7.3.– The previous calculation was conducted by taking u a solution of a problem in the family $(\mathcal{P}_{s,h})$, but formula [7.34] is written for a fixed value of s, as u and v are arbitrary test functions $H^{1/2}(\Gamma_s)$. If we now consider u and v as transported by the velocity field V of arbitrary functions in $H^{1/2}(\Gamma_0)$, we have, by means of equation [7.32] from [DEL 01]:

$$\frac{\mathrm{d}}{\mathrm{d}s}(Pu,v)_{\Gamma_s} = (\frac{\mathrm{d}P}{\mathrm{d}s}u,v)_{\Gamma_s} + (\beta Pu,v)_{\Gamma_s}, \qquad [7.37]$$

and so:

$$\begin{aligned} &\frac{\mathrm{d}}{\mathrm{d}s}(Pu,v)_{\Gamma_s} + (\alpha Pu,Pv)_{\Gamma_s} + (PV_{\Gamma_s}\cdot\nabla_{\Gamma_s}u,v)_{\Gamma_s} \\ &\quad + (u,PV_{\Gamma_s}\cdot\nabla_{\Gamma_s}v)_{\Gamma_s} = \int_{\Gamma_s}\alpha\nabla_{\Gamma_s}u\cdot\nabla_{\Gamma_s}v\,\mathrm{d}\Gamma_s, \\ &P(0)=0. \end{aligned} \qquad [7.38]$$

Let us show that formula [7.38] is compatible with self-adjointness of P. Indeed, we have:

$$\frac{\mathrm{d}}{\mathrm{d}s}(Pu,v)_{\Gamma_s} = \frac{\mathrm{d}}{\mathrm{d}s}(Pv,u)_{\Gamma_s} \qquad [7.39]$$

By integration w.r.t. s, this proves the self-adjointness of $P(s)$ because $P(0)=0$.

In addition, for a numerical approach of the type presented in section 5.6 but with surfaces Γ_{s_i} of different shapes and assuming that there is a one-to-one correspondence between the nodes of each surface, formulation [7.38] allows a simpler discretization than [7.34]: indeed, a finite difference discretization of the term $\frac{\mathrm{d}}{\mathrm{d}s}(Pu,v)_{\Gamma_s}$ is written simply using the shape functions of each of the surfaces Γ_{s_i} and $\Gamma_{s_{i+1}}$, and this avoids having to calculate β.

We now show that a similar calculation can be made for the backward invariant embedding from section 1.3, which provides an equation for the Neumann-to-Dirichlet operator. Therefore, we consider the problem:

$$(\widetilde{\mathcal{P}}_{s,h})\begin{cases} -\Delta u = f \quad \text{in } \hat{\Omega}_s, \\ \dfrac{\partial u}{\partial n}|_{\Gamma_s} = h \in H^{-1/2}(\Gamma_s), \quad u|_{\Gamma_1} = u_1 \in H^{1/2}(\Gamma_1), \end{cases} \qquad [7.40]$$

where $\hat{\Omega}_s$ is the open set limited by Γ_s and Γ_1, and therefore $\overline{\hat{\Omega}_s \bigcup \tilde{\Omega}_s} = \overline{\Omega}$. By the argument of invariant embedding of section 1.3, we have the relation:

$$(u|_{\Gamma_s}, v)_{\Gamma_s} = (Q(s)\frac{\partial u}{\partial n}|_{\Gamma_s}, v)_{\Gamma_s} + (r(s), v)_{\Gamma_s}, \quad \forall v \in H^{-1/2}(\Gamma_s), \quad [7.41]$$

where $Q(s)$ is the Neumann-to-Dirichlet operator on Γ_s, which is a positive and self-adjoint operator in $\mathcal{L}(H^{-1/2}(\Gamma_s), H^{1/2}(\Gamma_s))$ that vanishes on Γ_1.

Now, we take $Q(s)v$ as a test function, where v is an arbitrary function on Γ_s transported by the velocity field V as s varies, and we use Green's formula applied to problem [7.40] and $Q(1) = 0$:

$$\begin{aligned}\int_{\hat{\Omega}_s} \nabla u \cdot \nabla Q v \, \mathrm{d}x &= \int_{\hat{\Omega}_s} f Q v \, \mathrm{d}x + \int_{\Gamma_s} \frac{\partial u}{\partial n}|_{\Gamma_s} Q(s) v \, \mathrm{d}\Gamma_s \\ &= \int_{\hat{\Omega}_s} f Q v \, \mathrm{d}x + \int_{\Gamma_s} Q(s)\frac{\partial u}{\partial n}|_{\Gamma_s} v \, \mathrm{d}\Gamma_s \qquad [7.42] \\ &= \int_{\hat{\Omega}_s} f Q v \, \mathrm{d}x + \int_{\Gamma_s} (u|_{\Gamma_s} - r(s)) v \, \mathrm{d}\Gamma_s.\end{aligned}$$

The normal to Γ_s is now oriented outwards from $\hat{\Omega}_s$ and formula [7.21] becomes:

$$\frac{\partial u}{\partial s} = -\alpha \frac{\partial u}{\partial n} + V_{\Gamma_s} \cdot \nabla_{\Gamma_s} u. \qquad [7.43]$$

Returning to the calculations of formula [7.27], choosing v independent of s but taking into account that Q depends on s, we find:

$$\begin{aligned}\int_{\hat{\Omega}_{s_0}} \nabla u {\cdot} \nabla (Qv) \, \mathrm{d}x = \int_{s_0}^{1} \int_{\Gamma_s} \Big(&\frac{1}{\alpha^2} \frac{\partial u}{\partial s} \frac{\mathrm{d}Q}{\mathrm{d}s} v - \frac{1}{\alpha^2} (\frac{\partial u}{\partial s} V_{\Gamma_s} \cdot \nabla_{\Gamma_s}(Qv) \\ &+ \frac{\mathrm{d}Q}{\mathrm{d}s} v V_{\Gamma_s} \cdot \nabla_{\Gamma_s} u) + \frac{1}{\alpha^2} V_{\Gamma_s} \cdot \nabla_{\Gamma_s} u V_{\Gamma_s} \cdot \nabla_{\Gamma_s}(Qv) \\ &+ \nabla_{\Gamma_s} u \cdot \nabla_{\Gamma_s}(Qv) \Big) \alpha \, \mathrm{d}\Gamma_s \, \mathrm{d}s. \qquad [7.44]\end{aligned}$$

We can rewrite this formula:

$$\int_{\hat{\Omega}_{s_0}} \nabla u \cdot \nabla(Qv)\,\mathrm{d}x = \int_{s_0}^{1}\int_{\Gamma_s}\Big(-\frac{1}{\alpha}\frac{\partial u}{\partial n}\frac{\mathrm{d}Q}{\mathrm{d}s}v + \frac{1}{\alpha}\frac{\partial u}{\partial n}V_{\Gamma_s}\cdot\nabla_{\Gamma_s}(Qv) + \nabla_{\Gamma_s}u\cdot\nabla_{\Gamma_s}(Qv)\Big)\,\mathrm{d}\Gamma_s\,\mathrm{d}s. \quad [7.45]$$

Now, differentiating [7.42] with respect to s using [7.45] for the left-hand side, we obtain:

$$-\int_{\Gamma_s}\Big(-\frac{1}{\alpha}\frac{\partial u}{\partial n}\frac{\mathrm{d}Q}{\mathrm{d}s}v + \frac{1}{\alpha}\frac{\partial u}{\partial n}V_{\Gamma_s}\cdot\nabla_{\Gamma_s}(Qv) + \nabla_{\Gamma_s}u\cdot\nabla_{\Gamma_s}(Qv)\Big)\,\alpha\,\mathrm{d}\Gamma_s$$
$$= -\int_{\Gamma_s}\Big(-\frac{1}{\alpha}\frac{\partial u}{\partial n}\frac{\mathrm{d}Q}{\mathrm{d}s}v + \frac{1}{\alpha}\frac{\partial u}{\partial n}V_{\Gamma_s}\cdot\nabla_{\Gamma_s}(Qv) + \nabla_{\Gamma_s}(Q\frac{\partial u}{\partial n} + r)\cdot\nabla_{\Gamma_s}(Qv)\Big)\alpha\,\mathrm{d}\Gamma_s, \quad [7.46]$$

and for the derivative of the right-hand side:

$$-\int_{\Gamma_s} f|_{\Gamma_s}Q(s)v\alpha\,\mathrm{d}\Gamma_s + \int_{\Gamma_s}\frac{\partial}{\partial s}((u|_{\Gamma_s} - r(s))v)\,\mathrm{d}\Gamma_s + \int_{\Gamma_s}\beta(u|_{\Gamma_s} - r(s))v\,\mathrm{d}\Gamma_s$$
$$= -\int_{\Gamma_s} f|_{\Gamma_s}Q(s)v\alpha\,\mathrm{d}\Gamma_s + \int_{\Gamma_s}(-\alpha\frac{\partial u}{\partial n} + V_{\Gamma_s}\cdot\nabla_{\Gamma_s}(Q\frac{\partial u}{\partial n}) + \beta Q\frac{\partial u}{\partial n})v\,\mathrm{d}\Gamma_s + \int_{\Gamma_s}(-\frac{\partial r}{\partial s} + V_{\Gamma_s}\cdot\nabla_{\Gamma_s}r)v\,\mathrm{d}\Gamma_s. \quad [7.47]$$

By the reasoning of section 1.3, $\frac{\partial u}{\partial n} = w$ and v are arbitrary in $H^{-1/2}(\Gamma_s)$ and we deduce the Riccati equation for $Q(s)$:

$$(\frac{\mathrm{d}Q}{\mathrm{d}s}v, w)_{\Gamma_s} - (\alpha\nabla_{\Gamma_s}\circ Qw, \nabla_{\Gamma_s}\circ Qv)_{\Gamma_s} - (V_{\Gamma_s}\cdot\nabla_{\Gamma_s}\circ Qw, v)_{\Gamma_s} - (w, V_{\Gamma_s}\cdot\nabla_{\Gamma_s}\circ Qv)_{\Gamma_s} - (\beta Qw, v)_{\Gamma_s} + (\alpha w, v)_{\Gamma_s} = 0, \quad [7.48]$$

$$Q(1) = 0, \quad [7.49]$$

and the equation for the residual $r(s)$:

$$-(\frac{\partial r}{\partial s}, v)_{\Gamma_s} + (\alpha\nabla_{\Gamma_s} r, \nabla_{\Gamma_s} \circ Qv)_{\Gamma_s}$$

$$+ (V_{\Gamma_s} \cdot \nabla_{\Gamma_s} r, v)_{\Gamma_s} = (f, \alpha Qv)_{\Gamma_s}, \qquad [7.50]$$

$$r(1) = u_1. \qquad [7.51]$$

Once Q and r have been calculated, the state u is given by solving [7.41] for s ranging from 0 to 1, which can be rewritten as:

$$(Q\frac{1}{\alpha}\frac{\partial u}{\partial s}, v)_{\Gamma_s} + (u, v)_{\Gamma_s} - (Q\frac{1}{\alpha}V_{\Gamma_s} \cdot \nabla_{\Gamma_s} u, v)_{\Gamma_s} = (r, v)_{\Gamma_s}, \qquad [7.52]$$

$$u(0) = Q(0)u_0 + r(0). \qquad [7.53]$$

REMARK 7.4.– The verification that equation [7.49] is compatible with the self-adjointness of Q is similar to that of remark 7.3 but is a bit more complicated as the derivative:

$$\frac{\mathrm{d}}{\mathrm{d}s}(Qu, v)_{\Gamma_s} = (\frac{\mathrm{d}Q}{\mathrm{d}s}u, v)_{\Gamma_s} + (\beta Qu, v)_{\Gamma_s}, \qquad [7.54]$$

which holds for functions u and v transported by the V field, does not match the term that occurs in [7.49]. However, we can see from [7.49] that the quantity $(\frac{\mathrm{d}Q}{\mathrm{d}s}v, w)_{\Gamma_s} - (\beta Qw, v)_{\Gamma_s}$ is invariant under permutation of v and w. Therefore:

$$(\frac{\mathrm{d}Q}{\mathrm{d}s}v, w)_{\Gamma_s} - (\beta Qw, v)_{\Gamma_s} = (\frac{\mathrm{d}Q}{\mathrm{d}s}w, v)_{\Gamma_s} - (\beta Qv, w)_{\Gamma_s},$$

that is:

$$\frac{\mathrm{d}}{\mathrm{d}s}(Qv, w)_{\Gamma_s} = (\frac{\mathrm{d}Q}{\mathrm{d}s}v, w)_{\Gamma_s} + (\beta Qv, w)_{\Gamma_s}$$

$$= (\frac{\mathrm{d}Q}{\mathrm{d}s}w, v)_{\Gamma_s} + (\beta Qw, v)_{\Gamma_s} = \frac{\mathrm{d}}{\mathrm{d}s}(Qw, v)_{\Gamma_s}. \qquad [7.55]$$

We therefore deduce the self-adjointness of Q as in remark 7.3.

EXAMPLE 7.2.– *Homothetic domains*

Assume that Γ_0 and Γ_1 are homothetic in $\mathbb{R}^2$ with respect to $O \in \tilde{\Omega}_0$, which is taken as the origin of the coordinate system. We further assume that

$\tilde{\Omega}_0$ (and also consequently $\tilde{\Omega}_1$) is star shaped around 0. The family of curves $\{\Gamma_s\}_{s=0}^{s=1}$ used for the invariant embedding is defined with respect to Γ_0 by the homothety $\mathcal{H}(O,\lambda)$ with center O and ratio λ, which transform Γ_0 to Γ_s, such that the largest ratio λ_m corresponds to the transformation of Γ_0 to Γ_1. We obtain:

$$\Gamma_s = \mathcal{H}(O,\lambda)\Gamma_0, \quad \Gamma_1 = \mathcal{H}(O,\lambda_m)\Gamma_0 \text{ with } s = \frac{\lambda - 1}{\lambda_m - 1}.$$

Hence, $s = 1$ means Γ_1 and $s = 0$ means Γ_0.

Let t be the curvilinear abscissa along Γ_0, $\rho = \left\|\overrightarrow{OM}\right\|$ and θ be the angle $(\overrightarrow{OM}, \vec{n})$, where $\vec{n}$ is the outward normal to Ω_0. Let us remark that this angle is constant for all Γ_s along the line (OM) by the properties of the homothety.

We define the function $t \mapsto \varphi(t)$ on the curve Γ_0 by:

$$\Gamma_0 = \{(t,\rho) | \rho = \varphi(t)\}.$$

Assume that the curve Γ_0 is such that:

$$-\frac{\pi}{2} < \theta_- \leq \theta \leq \theta_+ < \frac{\pi}{2}.$$

The angle θ is related to the curvilinear abscissa t along Γ_0 by:

$$\frac{\mathrm{d}\varphi}{\mathrm{d}t} = \sin\theta.$$

As mentioned earlier, we take (s,t) as local coordinate system, that is, the second coordinate t at M is the curvilinear abscissa along Γ_0 of M' intersection of (OM) with Γ_0. We can then compute the functions that occur in formulas [7.34], [7.35] and [7.36]. We have:

$$V(M) = (\lambda_m - 1)\overrightarrow{OM'}, \quad V_{\Gamma_s} = (\lambda_m - 1)\sin\theta\,\varphi. \qquad [7.56]$$

By the homothety, we have:

$$\nabla_{\Gamma_s} = \frac{1}{\lambda}\frac{\mathrm{d}}{\mathrm{d}t}, \text{ and therefore, } V_{\Gamma_s} \cdot \nabla_{\Gamma_s} = (\lambda_m - 1)\sin\theta\,\frac{\varphi}{\lambda}\frac{\mathrm{d}}{\mathrm{d}t} \qquad [7.57]$$

$$\alpha = (\lambda_m - 1)\cos\theta\,\varphi. \qquad [7.58]$$

In order to compute β, let us remark that:

$$\lambda V = (\lambda_m - 1)\overrightarrow{OM}.$$

And therefore:

$$D(\lambda V) = \lambda DV + \nabla\lambda \otimes V = (\lambda_m - 1)I. \qquad [7.59]$$

Now:

$$\nabla\lambda = \frac{(\lambda_m - 1)}{\varphi\cos\theta}\,\vec{n}. \qquad [7.60]$$

By placing ourselves in a basis linked to the normal and the tangent to Γ_s, we find that:

$$\nabla\lambda \otimes V = (\lambda_m - 1)\begin{pmatrix} 1 & \mathrm{tg}\theta \\ 0 & 0 \end{pmatrix}, \qquad [7.61]$$

and so:

$$\lambda DV = (\lambda_m - 1)\begin{pmatrix} 0 & -\mathrm{tg}\theta \\ 0 & 1 \end{pmatrix}. \qquad [7.62]$$

Thus:

$$\lambda D(V)\vec{n}.\vec{n} = 0. \qquad [7.63]$$

Hence:

$$\beta = \mathrm{div}V = \frac{\lambda_m - 1}{\lambda}. \qquad [7.64]$$

REMARK 7.5.– All calculations naturally extend to dimensions greater than 2. However, note that formula [7.64] is limited to the dimension 2. For example, in 3D, we have:

$$\beta = \mathrm{div}V = 2\frac{\lambda_m - 1}{\lambda}. \qquad [7.65]$$

Formulas [7.34], [7.35] and [7.36] are written on Γ_s. Using homotheties, it is possible to reduce the integrals by a change in the variable to integrals along Γ_0 causing the corresponding factor to simplify. In this context, the factorized system reads:

$$\frac{1}{\lambda_m - 1}(\frac{\mathrm{d}P}{\mathrm{d}s}u, v)_{\Gamma_0} + (\varphi\cos\theta Pu, Pv)_{\Gamma_0} + (P\frac{\varphi}{\lambda}\sin\theta\frac{\partial u}{\partial t}, v)_{\Gamma_0}$$
$$+ (u, P\frac{\varphi}{\lambda}\sin\theta\frac{\mathrm{d}v}{\mathrm{d}t})_{\Gamma_0} + \frac{1}{\lambda}(Pu, v)_{\Gamma_0} = (\frac{\varphi}{\lambda^2}\cos\theta\frac{\mathrm{d}u}{\mathrm{d}t}, \frac{\mathrm{d}v}{\mathrm{d}t})_{\Gamma_0}. \quad [7.66]$$

This equation is to be integrated from $s = 0$ to $s = 1$, with $P(0) = 0$ and $\lambda = 1 + (\lambda_m - 1)s$. The same applies to the equation on the residual w:

$$\frac{1}{\lambda_m - 1}(\frac{\partial w}{\partial s}, v)_{\Gamma_0} + (P\varphi\cos\theta\, w, v)_{\Gamma_0} + (w, \frac{\varphi}{\lambda}\sin\theta\frac{\mathrm{d}v}{\mathrm{d}t})_{\Gamma_0} + \frac{1}{\lambda}(w, v)_{\Gamma_0}$$
$$+ (\varphi\cos\theta f, v)_{\Gamma_0} = 0, \quad [7.67]$$

with the initial condition:

$$w(0) = -u_0.$$

The solution u is then obtained by integrating from $s = 1$ to $s = 0$:

$$\frac{1}{\lambda_m - 1}(\frac{\partial u}{\partial s}, v)_{\Gamma_0} = (\varphi\cos\theta Pu, v)_{\Gamma_0} + (\frac{\varphi}{\lambda}\sin\theta\frac{\partial u}{\partial t}, v)_{\Gamma_0}$$
$$+ (\varphi\cos\theta w, v)_{\Gamma_0}, \quad [7.68]$$

with:

$$u(1) = u_1.$$

The function $u \in H^1(\Gamma_0)$ being arbitrary in formula [7.66] as well as $v \in H^1(\Gamma_0)$ in [7.66], [7.67] and [7.68], we have:

$$\frac{1}{\lambda_m - 1}\frac{\mathrm{d}P}{\mathrm{d}s} + \frac{1}{\lambda}P + P \circ \frac{\varphi}{\lambda}\sin\theta\frac{\partial}{\partial t} + (\frac{\varphi}{\lambda}\sin\theta\frac{\partial}{\partial t})^{\star} \circ P + P\varphi\cos\theta P$$
$$= -\frac{\partial}{\partial t}(\frac{\varphi}{\lambda^2}\cos\theta\frac{\partial}{\partial t}),$$

$$\frac{1}{\lambda_m - 1}\frac{\partial w}{\partial s} + \frac{1}{\lambda}w - \frac{\partial}{\partial t}(\frac{\varphi}{\lambda}\sin\theta w) + P\varphi\cos\theta\, w + \varphi\cos\theta f = 0,$$

$$\frac{1}{\lambda_m - 1}\frac{\partial u}{\partial s} = \varphi\cos\theta P u + \frac{\varphi}{\lambda}\sin\theta\frac{\partial u}{\partial t} + \varphi\cos\theta w.$$

By making the change of unknown:

$$P = \frac{P'}{\lambda}, \quad w = \frac{w'}{\lambda},$$

and of variable:

$$\mu = \log(\lambda),$$

we obtain a system whose coefficients are independent of s:

$$\frac{\mathrm{d}P'}{\mathrm{d}\mu} + P' \circ \varphi\sin\theta\frac{\partial}{\partial t} + (\varphi\sin\theta\frac{\partial}{\partial t})^\star \circ P' + P'\varphi\cos\theta P'$$

$$= -\frac{\partial}{\partial t}(\varphi\cos\theta\frac{\partial}{\partial t}), \quad [7.69]$$

$$\frac{\partial w'}{\partial \mu} - \frac{\partial}{\partial t}(\varphi\sin\theta w') + P'\varphi\cos\theta\, w' + \lambda^2\varphi\cos\theta f = 0, \quad [7.70]$$

$$\frac{\partial u}{\partial \mu} = \varphi\cos\theta P' u + \varphi\sin\theta\frac{\partial u}{\partial t} + \varphi\cos\theta w'. \quad [7.71]$$

The new independent variable μ varies from 0 to $\mu_m = \log(\lambda_m)$.

EXAMPLE 7.3.– *The computing zoom*

The principle of factorization by homothety presented above was used in [HEN 07] to define what was called the "computing zoom". The principle is the following: during a numerical simulation of the solution of a boundary value problem set in its natural domain, we might be particularly interested in computing this solution in a small area: the region of interest. For example, this region may contain abrupt changes in the coefficients, nonlinearities, model uncertainties, optimization (optimal control), etc., and so this may cause highly repetitive calculations. We might then wish to concentrate the computing power within the region of interest without neglecting the interactions outside this area.

Another situation is the one where there are real-time constraints on the calculation, as is the case for virtual reality, and where we therefore seek to minimize the region where the calculations are repeated: for example, for the calculation of a haptic feedback, it may be sufficient to calculate the forces and movements in a neighborhood of interaction, provided that their effect at a longer distance is not neglected.

The mathematical solution is to calculate the "transparent" boundary conditions along the region of interest, that is to say, the conditions which accurately reflect the effect on the solution of the domain outside the region of interest.

Assuming that the region of interest is a domain homothetic to the natural domain of the boundary value problem (or is a subset of such a domain), the invariant embedding by homothety presented above provides the solution.

We consider the following problem with Robin boundary conditions (these conditions ensure the coercivity of the boundary value problem):

$$(\mathcal{P}_0)\begin{cases} -\Delta u = f \quad \text{in } \Omega, f \in L^2(\Omega) \\ \dfrac{\partial u}{\partial n}|_\Gamma + ku|_\Gamma = u_0 \in H^{-1/2}(\Gamma), \quad k > 0. \end{cases}$$

Let Ω_λ be the image of Ω under the homothety $\mathcal{H}(0, \lambda)$ with center O in Ω and ratio $\lambda < 1$ and let Ω_{λ_0} be the region of interest. Then, we proceed to perform invariant embedding on the family of domains $\{\Omega \setminus \Omega_\lambda\}_{\lambda=\lambda_0}^{\lambda=1}$.The calculations leading to [7.69], [7.70] and [7.71] are still valid provided that the sign of λ is changed and λ is integrated from $\lambda = 1$ to $\lambda = \lambda_0$. Because of the Robin condition, the initial condition for the Riccati equation [7.69] becomes:

$$P'(0) = kI.$$

Then, once $P'(\mu_0)$ and $r'(\mu_0)$ have been calculated, the restriction of the solution of $(\mathcal{P}_0)$ to the region of interest Ω_{λ_0} is the solution of the problem:

$$(\mathcal{P}_{\lambda_0})\begin{cases} -\Delta u = f \quad \text{in } \Omega_{\lambda_0}, f \in L^2(\Omega_{\lambda_0}) \\ \dfrac{\partial u}{\partial n}|_{\Gamma_{\lambda_0}} = \lambda_0(P'u + r')|_{\Gamma_{\lambda_0}}, \quad k > 0. \end{cases}$$

The invariant embedding has eliminated the unknown u outside of Ω_{λ_0}. We can solve problem $(\mathcal{P}_{\lambda_0})$ by choosing the most appropriate numerical method. The calculation of P' is independent of the data f, u_0. If u_0 is changed or f is changed on $\Omega \setminus \Omega_{\lambda_0}$, it is necessary to recalculate r'. If after solving $(\mathcal{P}_{\lambda_0})$, we wish to have u on $\Omega \setminus \Omega_{\lambda_0}$, it may be obtained by integrating [7.71] starting from the value of u calculated on Γ_{λ_0}. Finally, note that if we are to increase the "zoom magnification", that is to say, reduce the size of Ω_{λ_0} by decreasing λ_0, it is sufficient to integrate further [7.69] from the value of P' already obtained.

REMARK 7.6.– Note that if we wished to apply this method of elimination by invariant embedding from the boundary Γ not just to the annulus limited by Γ and Γ_{λ_0}, but to the whole open set Ω with boundary Γ, there is a difficulty linked to the singularity at O. Indeed, there is a particular difficulty due to the definition of the initial condition for u at the point O, which is not given and which is generally not even defined. This problem has been studied by Maria do Céu Soares in her PhD thesis [SOA 06].

EXAMPLE 7.4.– Returning to the quasi-cylindrical domains discussed in section 7.2, we consider problem [7.5]. Suppose that, for any pair of sections, $\mathcal{O}_x$ and $\mathcal{O}_{x'}$ there is a $\mathcal{C}^1$ diffeomorphism from one to the other that also transforms $\partial\mathcal{O}_x$ into $\partial\mathcal{O}_{x'}$. The velocity field $V = \dfrac{\mathrm{d}\Phi_0^x}{\mathrm{d}x}$ on Σ is tangent to Σ. In the presentation of this section, it was assumed that the domains used for the invariant embedding had no boundaries. But, in the present situation, the deformation satisfies the $\mathcal{C}^1$ regularity necessary for the use of theorem 4.3 p. 355 [DEL 01]. Thus, the calculation leading to formulas [7.34], [7.35] and [7.36] still applies. As the section $\mathcal{O}_s$ is planar and moves parallel to itself, we can take $\alpha = 1$ and obtain:

$$\beta = \nabla_\Gamma \cdot V_\Gamma,$$

$$\left(\frac{\mathrm{d}P}{\mathrm{d}x}u, v\right)_\Gamma + (Pu, Pv)_\Gamma + (PV_\Gamma \cdot \nabla_\Gamma u, v)_\Gamma + (u, PV_\Gamma \cdot \nabla_\Gamma v)_\Gamma + (\beta P u, v)_\Gamma = \int_\Gamma \nabla_\Gamma u \cdot \nabla_\Gamma v \, \mathrm{d}\Gamma,$$

$$P(0) = 0, \qquad [7.72]$$

$$\left(\frac{\partial w}{\partial x}, v\right)_\Gamma + (Pw, v)_\Gamma + (w, V_\Gamma \cdot \nabla_\Gamma v)_\Gamma + (\beta w, v)_\Gamma + (f, v)_\Gamma = 0,$$

$$w(0) = -u_0, \qquad [7.73]$$

$$(\frac{\partial u}{\partial x}, v)_\Gamma = (Pu, v)_\Gamma + (V_\Gamma \cdot \nabla_\Gamma u, v)_\Gamma + (w, v)_\Gamma, \quad u(1) = u_1. \quad [7.74]$$

Together, formulas [7.9], [7.10], [7.7], [7.19] and [7.20], on the one hand, and [7.72], [7.73] and [7.74], on the other hand, provide two possible factorizations of problem [7.5], corresponding to two displacement fields of the mobile boundary of the invariant embedding. Note that the family of problems used for the invariant embedding is the same in both cases; they are the point displacement fields on the moving surface that are different.

8

Factorization by the QR Method

In Chapter 5, we saw that in the context of a method of finite differences, one particular discretization of the factorization [1.25] of problem $(\mathcal{P}_0)$ of section 1.1 corresponds exactly to the block Gaussian LU factorization of the matrix of the discretized problem $(\mathcal{P}_0)$. In other words, this factorization [1.25] may be viewed as an infinite-dimensional generalization of the Gaussian factorization. It is natural to ask ourselves if there also exists a generalization of the QR method, where Q is an orthogonal matrix and R is an upper triangular matrix. We know that the QR method applies not only to square matrices, but also to rectangular matrices with more rows than columns and so corresponds to overdetermined problems that we can solve in the sense of least squares. This generalization to overdetermined problems is also possible for the factorization of the boundary value problems studied here (this is discussed in the doctoral thesis of Maria Orey [ORE 12] and an additional boundary condition is discussed in [HEN 10]), but we will restrict our attention in this chapter to well-posed problems.

8.1. Normal equation for problem $(\mathcal{P}_0)$ in section 1.1

We begin by recalling a classic property of QR factorization in linear algebra. Let A be a non-singular $(n \times n)$ matrix and suppose that we wish to solve the linear system:

$$Ax = b, \quad x, b \in \mathbb{R}^n. \qquad [8.1]$$

The normal equation of this system is:

$$A^T A x = A^T b. \qquad [8.2]$$

The matrix $A^T A$ is symmetric positive definite. We can apply the Cholesky factorization to obtain:

$$A^T A = B^T B, \qquad [8.3]$$

where B is an $(n \times n)$ upper triangular matrix with positive entries on the diagonal. Multiplying [8.3] by A^{-T}, the inverse of A^T, we obtain:

$$A = A^{-T} B^T B = QB, \qquad [8.4]$$

setting $Q = A^{-T} B^T$. We will show that this matrix is orthogonal. Indeed:

$$QQ^T = A^{-T} B^T B A^{-1} = A^{-T} A^T A A^{-1} = I. \qquad [8.5]$$

Therefore, formula [8.4] corresponds to the factorization of A into the product of an orthogonal matrix and an upper triangular matrix with positive entries on the diagonal.

We will proceed similarly for problem $(\mathcal{P}_0)$, setting $a = 1$ for simplicity. In order to write the normal equation for $(\mathcal{P}_0)$, we will adopt additional regularity hypotheses on the data. We assume that:

$$f \in X \text{ as in definition 1.1}, \qquad [8.6]$$

$$u_0 \in H^{1/2}_{0,0}(\mathcal{O}), \qquad [8.7]$$

$$u_1 \in H^{3/2}(\mathcal{O}) \cap H^1_0(\mathcal{O}), \qquad [8.8]$$

or alternatively, using the hypotheses and notation of definitions 2.1 and 2.2:

$$f \in X_a(1), \qquad [8.9]$$

$$u_0 \in H^{1/2}, \qquad [8.10]$$

$$u_1 \in H^{3/2}. \qquad [8.11]$$

This implies in particular that f is zero on Σ.

As the operator associated with $(\mathcal{P}_0)$ is self-adjoint, the normal equation associated with this problem is:

$$\begin{cases} \Delta^2 u = -\Delta f \quad \text{in } \Omega, \\ u|_\Sigma = \Delta u|_\Sigma = 0, \quad -\dfrac{\partial u}{\partial x}|_{\Gamma_0} = u_0, \ \dfrac{\partial \Delta u}{\partial x}|_{\Gamma_0} = -\dfrac{\partial f}{\partial x}|_{\Gamma_0} \\ u|_{\Gamma_1} = u_1, \ \Delta u|_{\Gamma_1} = -f|_{\Gamma_1}. \end{cases} \qquad [8.12]$$

Under the hypotheses [8.9], [8.10] and [8.11], $(\mathcal{P}_0)$ and [8.12] have the same solution in $X_a(2)$.

8.2. Factorization of the normal equation by invariant embedding

We proceed as in section 1.2, embedding problem [8.12] into a family of similar problems defined in Ω_s with an additional boundary condition on Γ_s:

$$u|_{\Gamma_s} = h \in H^{3/2}, \Delta u|_{\Gamma_s} = k \in H^{-1/2} = (H^{1/2}_{0,0}(\mathcal{O}))'.$$

This problem is well-posed, and we know that we can define the normal derivative of the solution on Γ_s in $H^{1/2}_{0,0}(\mathcal{O})$, setting:

$$\frac{\partial u}{\partial x}|_{\Gamma_s} = P(s)h + Q(s)k + r(s). \qquad [8.13]$$

The operators P and Q are continuous:

$$P(s) \in \mathcal{L}(H^{3/2}, H^{1/2}), \quad Q(s) \in \mathcal{L}(H^{-1/2}, H^{1/2}).$$

If we choose $k = -f|_{\Gamma_s}$, problem [8.12] naturally factorizes into:

$$-\Delta z = -\Delta f \quad \text{in } \Omega_s, \qquad [8.14]$$

$$z|_\Sigma = 0, \qquad [8.15]$$

$$\frac{\partial z}{\partial x}|_{\Gamma_0} = \frac{\partial f}{\partial x}|_{\Gamma_0}, \quad z|_{\Gamma_s} = f|_{\Gamma_s}. \qquad [8.16]$$

$$-\Delta u = z \quad \text{in } \Omega_s, \tag{8.17}$$

$$u|_{\Sigma} = 0, \tag{8.18}$$

$$-\frac{\partial u}{\partial x}|_{\Gamma_0} = u_0, \quad u|_{\Gamma_s} = h. \tag{8.19}$$

From [8.14], we deduce that $z = f$ and u is the solution of $(\mathcal{P}_{s,h})$. Thus, the family of problems considered in the invariant embedding contains problems $(\mathcal{P}_{s,h})$, but is in fact larger. The identity [8.13] may be written as:

$$\frac{\partial u}{\partial x}|_{\Gamma_s} = P(s)h - Q(s)f + r(s), \tag{8.20}$$

so P is the Dirichlet-to-Neumann operator defined in section 1.2, satisfying [1.22]. We also deduce that $w = r - Qf$ satisfies [1.23].

In the case where $k \neq -f|_{\Gamma_s}$, [8.12] decomposes into:

$$-\Delta z = -\Delta f \quad \text{in } \Omega_s, \tag{8.21}$$

$$z|_{\Sigma} = 0, \tag{8.22}$$

$$\frac{\partial z}{\partial x}|_{\Gamma_0} = \frac{\partial f}{\partial x}|_{\Gamma_0}, \quad z|_{\Gamma_s} = -k, \tag{8.23}$$

and u satisfies [8.17].

By applying the factorization method to this problem, with $z = -\Delta u$, we obtain that:

$$\begin{cases} \dfrac{\mathrm{d}t}{\mathrm{d}x} + Pt = -\Delta f, & t(0) = -\dfrac{\partial f}{\partial x}|_{\Gamma_0}, \\ -\dfrac{\mathrm{d}\Delta u}{\mathrm{d}x} + P\Delta u = -t, & \Delta u(1) = -f(1). \end{cases} \tag{8.24}$$

To obtain the equations satisfied by Q and r, we proceed as in section 1.2 by differentiating w.r.t. x the identity [8.13] along a trajectory where $h = u$, $k = \Delta u$. We find:

$$\frac{\partial^2 u}{\partial x^2}(x) = \frac{\mathrm{d}P}{\mathrm{d}x}(x)u(x) + P(x)\frac{\mathrm{d}u}{\mathrm{d}x}(x) + \frac{\mathrm{d}Q}{\mathrm{d}x}(x)\Delta u(x) \\ + Q(x)\frac{\mathrm{d}\Delta u}{\mathrm{d}x}(x) + \frac{\mathrm{d}r}{\mathrm{d}x}. \tag{8.25}$$

We can then replace $\frac{\partial u}{\partial x}$ in [8.13] and $\frac{\partial \Delta u}{\partial x}$ in the second equation of [8.24]:

$$\Delta u - \Delta_y u = \frac{dP}{dx}u + P(Pu + Q\Delta u + r) + \frac{dQ}{dx}\Delta u + Q(P\Delta u + t) + \frac{dr}{dx}. \quad [8.26]$$

As u and Δu are arbitrary in the identity [8.26], the operators applied to them are zero. We therefore recover the Riccati equation for P, and, for Q, we find a Lyapunov equation:

$$\frac{dP}{dx} + P^2 + \Delta_y = 0, \quad P(0) = 0, \quad [8.27]$$

$$\frac{dQ}{dx} + PQ + QP = I, \quad Q(0) = 0. \quad [8.28]$$

The initial conditions on P and Q result from the boundary conditions on Γ_0, [8.12] and [8.13]. The operators P and Q are self-adjoint and positive for $x > 0$. The term independent of u and Δu gives:

$$\frac{dr}{dx} + Pr + Qt = 0. \quad [8.29]$$

The initial condition for r is obtained from [8.13] and [8.19]:

$$r(0) = -u_0. \quad [8.30]$$

In summary, regrouping these results, we arrive at the decoupled form of the normal equation [8.12]:

$$\frac{dP}{dx} + P^2 + \Delta_y = 0, \quad P(0) = 0, \quad [8.31]$$

$$\frac{dQ}{dx} + PQ + QP = I, \quad Q(0) = 0, \quad [8.32]$$

$$\frac{dt}{dx} + Pt = -\Delta f, \quad t(0) = -\frac{\partial f}{\partial x}|_{\Gamma_0}, \quad [8.33]$$

$$\frac{dr}{dx} + Pr = -Qt, \quad r(0) = -u_0, \quad [8.34]$$

$$-\frac{\mathrm{d}\Delta u}{\mathrm{d}x} + P\Delta u = -t, \quad \Delta u(1) = -f(1), \tag{8.35}$$

$$-\frac{\mathrm{d}u}{\mathrm{d}x} + Pu = -Q\Delta u - r, \quad u(1) = u_1. \tag{8.36}$$

As for factorization [1.22], [1.23] and [1.24] of problem $(\mathcal{P}_0)$, in order to factorize the normal equation, we first solve equations [8.31] and [8.32] on P and Q in the direction of increasing x and [8.33] and [8.34] for the residuals r and t. We then solve [8.35] and [8.36] in the direction of decreasing x to obtain Δu and u.

Factorization [1.22], [1.23] and [1.24] may be easily recovered from this system. Indeed, it is sufficient to replace $-\Delta u$ in [8.36] by the given f. We must then calculate $r - Qf$, which as we saw earlier is a solution of [1.23]. Thus, with this approach, it is not necessary to solve [8.32], [8.33] and [8.35]. However, numerically solving equations [8.31] to [8.36] could possibly be useful as a test of the precision of the solution, for example, in order to adapt the grid in the x-direction.

We will now demonstrate how to write [8.31] to [8.36] in a factorized form using second-order operators in x, by eliminating the variables Δu and t. By eliminating t from [8.33] and [8.34], noting that Q is invertible for $x \neq 0$, we have the following, with the notation of remark 1.7:

$$\left(\frac{\mathrm{d}}{\mathrm{d}x} + P\right) \circ Q^{-1} \circ \left(\frac{\mathrm{d}}{\mathrm{d}x} + P\right) r = \Delta f. \tag{8.37}$$

Extracting Δu from [8.36]:

$$Q^{-1}\left(\frac{\mathrm{d}}{\mathrm{d}x} - P\right) u = \Delta u + Q^{-1}r, \tag{8.38}$$

and using [8.35]:

$$\left(\frac{\mathrm{d}}{\mathrm{d}x} - P\right) \circ Q^{-1} \circ \left(\frac{\mathrm{d}}{\mathrm{d}x} - P\right) u = t + \left(\frac{\mathrm{d}}{\mathrm{d}x} - P\right) \circ Q^{-1}r. \tag{8.39}$$

Eliminating t from [8.34]:

$$\left(\frac{\mathrm{d}}{\mathrm{d}x} - P\right) \circ Q^{-1} \circ \left(\frac{\mathrm{d}}{\mathrm{d}x} - P\right) u = \left(-Q^{-1} \circ \left(\frac{\mathrm{d}}{\mathrm{d}x} + P\right) \right.$$
$$\left. + \left(\frac{\mathrm{d}}{\mathrm{d}x} - P\right) \circ Q^{-1}\right) r. \quad [8.40]$$

Left-multiplying by Q:

$$Q \circ \left(\frac{\mathrm{d}}{\mathrm{d}x} - P\right) \circ Q^{-1} \circ \left(\frac{\mathrm{d}}{\mathrm{d}x} - P\right) u = \left(-\left(\frac{\mathrm{d}}{\mathrm{d}x} + P\right) \circ Q \right.$$
$$\left. + Q \circ \left(\frac{\mathrm{d}}{\mathrm{d}x} - P\right)\right) \circ Q^{-1} r. \quad [8.41]$$

But from [8.32], we have that:

$$-\left(\frac{\mathrm{d}}{\mathrm{d}x} + P\right) \circ Q + Q \circ \left(\frac{\mathrm{d}}{\mathrm{d}x} - P\right) = -I. \quad [8.42]$$

We therefore have that:

$$Q^2 \left(-\frac{\mathrm{d}}{\mathrm{d}x} + P\right) \circ Q^{-1} \circ \left(-\frac{\mathrm{d}}{\mathrm{d}x} + P\right) u = -r. \quad [8.43]$$

Finally, using [8.37]:

$$\left(\frac{\mathrm{d}}{\mathrm{d}x} + P\right) \circ Q^{-1} \circ \left(\frac{\mathrm{d}}{\mathrm{d}x} + P\right) \circ Q^2 \circ \left(-\frac{\mathrm{d}}{\mathrm{d}x} + P\right)$$
$$\circ Q^{-1} \circ \left(-\frac{\mathrm{d}}{\mathrm{d}x} + P\right) u = -\Delta f. \quad [8.44]$$

Setting:

$$\mathcal{R} = Q\left(-\frac{\mathrm{d}}{\mathrm{d}x} + P\right) \circ Q^{-1} \circ \left(-\frac{\mathrm{d}}{\mathrm{d}x} + P\right), \quad [8.45]$$

we hence obtain a factorization of the abstract operator appearing in the normal equation [8.12] in the form $\mathcal{R}^*\mathcal{R}$. The operator $\mathcal{R}$ corresponds to the Cauchy problem [8.35] and [8.36] in the direction of decreasing x. It may be viewed as a generalization of an upper triangular matrix. We set $A = -\Delta$ with the boundary conditions of problem $(\mathcal{P}_0)$, so that:

$$Au = \begin{pmatrix} -\Delta u \\ -\dfrac{\partial u}{\partial x}(0) \\ u(a) \end{pmatrix}.$$

We also set:

$$\mathcal{Q} = A^{-1}\mathcal{R}^*. \qquad [8.46]$$

Then, we have that:

$$A^*A = A^2 = \mathcal{R}^*\mathcal{R}, \qquad [8.47]$$

and, hence:

$$A = \mathcal{Q}\mathcal{R}. \qquad [8.48]$$

We show that $\mathcal{Q}$ is orthogonal. Let:

$$\mathfrak{f} = \begin{pmatrix} f \\ u_0 \\ u_1 \end{pmatrix}$$

be the vector of problem data. We have:

$$\mathcal{Q}\mathcal{Q}^*\mathfrak{f} = A^{-1}\mathcal{R}^*\mathcal{R}A^{-1}\mathfrak{f} = A^{-1}\mathcal{R}^*\mathcal{R}u = Au = \mathfrak{f}. \qquad [8.49]$$

Similarly:

$$\mathcal{Q}^*\mathcal{Q} = \mathcal{R}A^{-2}\mathcal{R}^* = \mathcal{R}(\mathcal{R}^*\mathcal{R})^{-1}\mathcal{R}^* = I. \qquad [8.50]$$

8.3. The QR method

We will now make explicit the operator $\mathcal{Q}$, which will allow us to repeat the calculation in section 8.2 at a more elementary level, as it is necessary for numerical approaches. To maintain the symmetry of the calculations, we begin by setting:

$$\tilde{r}(x) = Q^{-1}(x)r(x) \quad 0 < x \leq a. \tag{8.51}$$

As we noted earlier after equation [8.20], $Q(\tilde{r} - f)$ satisfies equation [1.23]:

$$(\frac{\mathrm{d}}{\mathrm{d}x} + P)Q(\tilde{r} - f) = -f. \tag{8.52}$$

the equation satisfied by Q:

$$\begin{aligned}(\frac{\mathrm{d}}{\mathrm{d}x} + P)Q\tilde{r} &= (\frac{\mathrm{d}}{\mathrm{d}x} + P)Qf - f \\ &= Q\frac{\mathrm{d}}{\mathrm{d}x}f + PQf + (I - PQ - QP)f - f \\ &= Q(\frac{\mathrm{d}}{\mathrm{d}x} - P)f.\end{aligned} \tag{8.53}$$

We can therefore define the operator $\mathcal{Q}^{-1}$, the inverse of $\mathcal{Q}$, as the composition of the following mappings:

$$\begin{pmatrix} f \\ u_0 \end{pmatrix} \mapsto r \mapsto \tilde{r}, \tag{8.54}$$

where the first step consists of applying $Q(\frac{\mathrm{d}}{\mathrm{d}x} - P)$ to f:

$$g = Q(\frac{\mathrm{d}}{\mathrm{d}x} - P)f, \tag{8.55}$$

then solving:

$$(\frac{\mathrm{d}}{\mathrm{d}x} + P)r = g, \quad r(0) = -u_0. \tag{8.56}$$

The second mapping in [8.54] is multiplication by Q^{-1}. have the following explicit formula for $\tilde{r}$, which satisfies [8.56], by using the evolution operator $R(s,x)$ defined in [3.43]:

$$\begin{aligned} r(x) &= R(0,x)u_0 - \int_0^x R(s,x)g(s)\,\mathrm{d}s \\ &= R(0,x)u_0 - \int_0^x R(s,x)(Q(s)(\frac{\mathrm{d}}{\mathrm{d}x} - P)f(s))\,\mathrm{d}s \\ &= R(0,x)u_0 + Q(x)f(x) + \int_0^x \frac{\partial}{\partial s}(R(s,x)Q(s))f(s)\,\mathrm{d}s \\ &+ \int_0^x R(s,x)Q(s)P(s)f(s)\,\mathrm{d}s. \end{aligned} \quad [8.57]$$

Now, from relation [3.48], and [8.32]:

$$\begin{aligned} \int_0^x \frac{\partial}{\partial s}(R(s,x)Q(s))f(s)\,\mathrm{d}s &= \int_0^x \frac{\partial}{\partial s}R(s,x)Q(s)f(s)\,\mathrm{d}s \\ &+ \int_0^x R(s,x)\frac{\mathrm{d}Q}{\mathrm{d}s}(s)f(s)\,\mathrm{d}s \\ &= \int_0^x R(s,x)P(s)Q(s)f(s)\,\mathrm{d}s \\ &+ \int_0^x R(s,x)(I - Q(s)P(s) - P(s)Q(s))f(s)\,\mathrm{d}s \\ &= \int_0^x R(s,x)f(s)\,\mathrm{d}s - \int_0^x R(s,x)Q(s)P(s)f(s)\,\mathrm{d}s. \end{aligned} \quad [8.58]$$

Overall, we find the following explicit expression for $\mathcal{Q}^{-1}$:

$$\tilde{r}(x) = Q^{-1}(x)R(0,x)u_0 + f(x) + Q^{-1}(x)\int_0^x R(s,x)f(s)\,\mathrm{d}s. \quad [8.59]$$

In order to show the orthogonality of $\mathcal{Q}$, we will perform a similar calculation on the adjoint $\mathcal{Q}^\star$ of $\mathcal{Q}$. Formally, if $\mathcal{Q}$ is defined based on [8.53] by:

$$(\frac{\mathrm{d}}{\mathrm{d}x} - P)^{-1}Q^{-1}(\frac{\mathrm{d}}{\mathrm{d}x} + P)Q, \quad [8.60]$$

then for $\mathcal{Q}^\star$, we have that:

$$Q(\frac{\mathrm{d}}{\mathrm{d}x} - P)Q^{-1}(\frac{\mathrm{d}}{\mathrm{d}x} + P)^{-1}. \qquad [8.61]$$

More precisely, to calculate the inverse of $\frac{\mathrm{d}}{\mathrm{d}x} + P$, we need an initial condition to be given. Thus, $\mathcal{Q}^\star$ is obtained as the composition of the two mappings:

$$\begin{pmatrix} f \\ u_0 \end{pmatrix} \mapsto g' \mapsto \tilde{r}', \qquad [8.62]$$

where g' satisfies:

$$\frac{\mathrm{d}g'}{\mathrm{d}x}(x) + P(x)g'(x) = f(x), \quad g'(0) = -v_0, \qquad [8.63]$$

with v_0 to be determined at a later point. We have:

$$g'(x) = R(0,x)v_0 - \int_0^x R(s,x)f(s)\,\mathrm{d}s. \qquad [8.64]$$

Then, $\tilde{r}'$ is given by:

$$\tilde{r}' = Q(x)(\frac{\mathrm{d}}{\mathrm{d}x} - P(x))Q^{-1}g'(x). \qquad [8.65]$$

We have the following:

$$\begin{aligned}
(\frac{\mathrm{d}}{\mathrm{d}x} - P(x))Q^{-1}g'(x) &= -Q^{-1}\frac{\mathrm{d}Q}{\mathrm{d}x}Q^{-1}g' + Q^{-1}\frac{\mathrm{d}}{\mathrm{d}x}g' - PQ^{-1}g' \\
&= (PQ^{-1} + Q^{-1}P - Q^{-2})g' + Q^{-1}\frac{\mathrm{d}}{\mathrm{d}x}g' - PQ^{-1}g' \\
&= (Q^{-1}P - Q^{-2})(R(0,x)v_0 - \int_0^x R(s,x)f(s)\,\mathrm{d}s) \\
&\quad + Q^{-1}(-P(x)R(0,x)v_0 + f(x) + \int_0^x P(x)R(s,x)f(s)\,\mathrm{d}s) \\
&= Q^{-1}f(x) - Q^{-2}R(0,x)v_0 + Q^{-2}\int_0^x R(s,x)f(s)\,\mathrm{d}s. \qquad [8.66]
\end{aligned}$$

$$\tilde{r}' = f(x) - Q^{-1}R(0,x)v_0 + Q^{-1}\int_0^x R(s,x)f(s)\,\mathrm{d}s. \qquad [8.67]$$

To identify this formula with [8.59], v_0 must be determined first. We want $\tilde{r}'$ to satisfy the same initial condition as $\tilde{r}$, so that:

$$\lim_{x\to 0} Q(x)\tilde{r}' = \lim_{x\to 0} Q(x)\tilde{r}, \qquad [8.68]$$

and hence from [8.59] and [8.67]:

$$v_0 = -u_0. \qquad [8.69]$$

With this choice, $\mathcal{Q}^{-1}$ and $\mathcal{Q}^{\star}$ are identical and so $\mathcal{Q}$ is indeed orthogonal. It should be noted that by means of this calculation, it is possible to obtain the operator $\mathcal{Q}$ and its adjoint using only first-order differentiation (and integration) w.r.t. x, whereas working from the definition [8.46] requires second-order derivatives. Implementing the QR method does not require the use of normal equations [8.12], which here simply served as an intermediate step in the calculation. We recover a well-known fact in numerical matrix analysis, namely that the QR methods (Householder or Givens) consist of the step-by-step construction of an orthogonal matrix as a product of elementary orthogonal matrices, but we do not reformulate the normal equation, as doing so would degrade the problem conditioning. We can draw a connection between these elementary transformations and relation [8.61], which produces an orthogonal relation between f and $\tilde{r}$ that is local in x.

The implementation of the QR method to solve problem $(\mathcal{P}_0)$ of section 1.1 may be summarized as follows:

– begin by calculating the operators P and Q, respectively, the solutions of [8.27] and [8.28], then the evolution operator $R(t,x)$, a solution of [3.43];

– from the data u_0, f and the orthogonal operator $\mathcal{Q}^{\star}$ (which is not calculated explicitly as we only need to use Q satisfying [8.28] and the evolution operator $R(t,x)$), we calculate $\tilde{r}(x)$ as follows:

$$\tilde{r}(x) = f(x) + Q^{-1}R(0,x)u_0 + Q^{-1}\int_0^x R(s,x)f(s)\,\mathrm{d}s,$$

for x ranging from 0 to 1;

– or alternatively, we can choose not to calculate $R(t,x)$, instead obtaining $\tilde{r}$ by solving:

$$(\frac{\mathrm{d}}{\mathrm{d}x} + P)r = Q(\frac{\mathrm{d}}{\mathrm{d}x} - P)f, \quad r(0) = -u_0, \tag{8.70}$$

so that $\tilde{r} = Q^{-1}r$;

– we then obtain the solution u by integrating the second-order differential equation in x from $x = 1$ to $x = 0$:

$$Q\left(-\frac{\mathrm{d}}{\mathrm{d}x} + P\right)Q^{-1}\left(-\frac{\mathrm{d}}{\mathrm{d}x} + P\right)u = -\tilde{r}, \tag{8.71}$$

$$u(1) = u_1, \quad \frac{\mathrm{d}u}{\mathrm{d}x}(1) = P(1)u_1 - Q(1)(f(1) - \tilde{r}(1)). \tag{8.72}$$

9

Representation Formulas for Solutions of Riccati Equations

In this chapter, we will apply some of the classical results on the Riccati equations to the Riccati equations encountered during applications of the factorization method for boundary value problems, reducing these equations to a linear form by means of a homography. In the model case of problem $(\mathcal{P}_0)$ in section 1.1, in which the Riccati equation has constant coefficients, this allows to find an explicit solution of the equation.

9.1. Representation formulas

Consider the mappings $X(x) \in \mathcal{L}(H_{00}^{1/2}(\mathcal{O}), H_{00}^{1/2}(\mathcal{O})')$ and $Y(x) \in \mathcal{L}(H_{00}^{1/2}(\mathcal{O}), H_{00}^{1/2}(\mathcal{O}))$ that send the Dirichlet condition of problem $(\mathcal{P}_0)$ on Γ_a to the Neumann and Dirichlet conditions on Γ_x:

$$X(x): \quad u(a) \mapsto \frac{\partial u}{\partial x}(x), \qquad [9.1]$$

$$Y(x): \quad u(a) \mapsto u(x), \qquad [9.2]$$

where u is the solution of $(\mathcal{P}_0)$ in the homogeneous case, that is, when $f = 0$ and $u_0 = 0$. The operator Y clearly satisfies $Y(a) = I$ and it follows from the Neumann condition of problem $(\mathcal{P}_0)$ that $X(0) = 0$. The continuity of these mappings follows from the well-posedness of $(\mathcal{P}_0)$, and an application of proposition 1.1. However, as $u(a)$ ranges over $H_{00}^{1/2}(\mathcal{O})$, u does not range over the full space X defined in definition 1.1 since $f = 0$ and so Y is not surjective in $H_{00}^{1/2}(\mathcal{O})$. From this, we have the following controllability result:

PROPOSITION 9.1.– The image of $H_{00}^{1/2}(\mathcal{O})$ under Y is dense in $H_{00}^{1/2}(\mathcal{O})$.

PROOF.– By the usual linear controllability arguments: if the image of Y is not dense in $H_{00}^{1/2}(\mathcal{O})$, there exists $\eta \in H_{00}^{1/2}(\mathcal{O})'$, $\eta \neq 0$, such that:

$$< \eta, u(x; u(a)) >_{H_{00}^{1/2}(\mathcal{O})' \times H_{00}^{1/2}(\mathcal{O})} = 0, \quad \forall u(a) \in H_{00}^{1/2}(\mathcal{O}), \qquad [9.3]$$

where $u(x; u(a))$ is the trace on Γ_x of the solution of $(\mathcal{P}_0)$ for $u_1 = u(a)$. Let p be the solution to:

$$\begin{cases} -\Delta p = 0 \quad \text{in } \widetilde{\Omega}_x =]x, a[\times \mathcal{O}, \\ p|_\Sigma = 0, \quad -\dfrac{\partial p}{\partial x}|_{\Gamma_x} = \eta, \quad p|_{\Gamma_a} = 0. \end{cases} \qquad [9.4]$$

By Green's formula:

$$\begin{aligned} &< \eta, u(x; u(a)) >_{H_{00}^{1/2}(\mathcal{O})' \times H_{00}^{1/2}(\mathcal{O})} \\ &= < \frac{\partial p}{\partial x}|_{\Gamma_a}, u(a) >_{H_{00}^{1/2}(\mathcal{O})' \times H_{00}^{1/2}(\mathcal{O})} = 0. \end{aligned} \qquad [9.5]$$

Then, $\dfrac{\partial p}{\partial x}|_{\Gamma_a}$ is zero and for p we have homogeneous Cauchy conditions on Γ_a. By Holmgren's theorem, p is zero and so is η, which contradicts the hypothesis and proves the result. □

Differentiating [9.1]:

$$\frac{\mathrm{d}}{\mathrm{d}x} X(x) u(a) = \frac{\partial^2 u}{\partial x^2}(x) = -\Delta_y u(x) = -\Delta_y Y(x) u(a). \qquad [9.6]$$

Thus, X and Y satisfy:

$$\begin{pmatrix} \dfrac{\mathrm{d}X}{\mathrm{d}x} \\ \dfrac{\mathrm{d}Y}{\mathrm{d}x} \end{pmatrix} = \begin{pmatrix} 0 & -\Delta_y \\ I & 0 \end{pmatrix} \begin{pmatrix} X \\ Y \end{pmatrix}, \qquad [9.7]$$

where $Y(a) = I$ and $X(0) = 0$.

Let P be defined by:

$$P(x) = X(x)Y(x)^{-1}. \qquad [9.8]$$

A priori, this operator is defined only on the image of Y. With this definition, it is clear from the definitions [9.1], [9.2] that it corresponds to the Dirichlet-to-Neumann operator from definition 1.2. By proposition 9.1, it can be extended by continuity to $H_{00}^{1/2}(\mathcal{O})$.

Formally differentiating [9.8]:

$$\begin{aligned}\frac{\mathrm{d}P}{\mathrm{d}x}(x) &= \frac{\mathrm{d}X}{\mathrm{d}x}(x)Y(x)^{-1} - X(x)Y(x)^{-1}\frac{\mathrm{d}Y}{\mathrm{d}x}Y(x)^{-1}\\ &= -\Delta_y - P(x)^2,\end{aligned}$$

we check that P is a solution of the Riccati equation with the initial condition:

$$P(0) = 0. \qquad [9.9]$$

We now wish to study the coupled system [9.7], which is a two-point boundary value problem. We will show that this system can be diagonalized.

9.2. Diagonalization of the two-point boundary value problem

Let P_0 be the positive solution of the algebraic Riccati equation:

$$P_0^2 = -\Delta_y, \qquad P_0 = (-\Delta_y)^{1/2}, \qquad P_0 \in \mathcal{L}(H_{00}^{1/2}(\mathcal{O}), H_{00}^{1/2}(\mathcal{O})'). \quad [9.10]$$

This operator corresponds to the asymptotic value, as $x \longrightarrow \infty$, of the Dirichlet-to-Neumann operator from definition 1.2. We introduce the operators $\Phi \in \mathcal{L}(H_{00}^{1/2}(\mathcal{O}), H_{00}^{1/2}(\mathcal{O})')$ and $\Psi \in \mathcal{L}(H_{00}^{1/2}(\mathcal{O}), H_{00}^{1/2}(\mathcal{O}))$, which become the new unknowns after performing the change of variables:

$$\begin{pmatrix} X \\ Y \end{pmatrix} = \begin{pmatrix} I & -P_0 \\ P_0^{-1} & I \end{pmatrix} \begin{pmatrix} \Phi \\ \Psi \end{pmatrix}. \qquad [9.11]$$

THEOREM 9.1.– The linear system:

$$\begin{cases} \dfrac{dX}{dx}(x) = -\Delta_y Y(x), \quad X(0) = 0, \\ \dfrac{dY}{dx}(x) = X(x), \quad Y(a) = I, \end{cases} \qquad [9.12]$$

is equivalent to the diagonal system:

$$\begin{cases} \dfrac{d\Phi}{dx}(x) = P_0\Phi(x), \quad \Phi(0) - P_0\Psi(0) = 0, \\ \dfrac{d\Psi}{dx}(x) = -P_0\Psi(x), \quad P_0^{-1}\Phi(a) + \Psi(a) = I, \end{cases} \qquad [9.13]$$

where X and Y are linked to Φ and Ψ by [9.11].

PROOF.– System [9.11] can be inverted to give:

$$\begin{pmatrix} \Phi \\ \Psi \end{pmatrix} = \frac{1}{2}\begin{pmatrix} I & P_0 \\ -P_0^{-1} & I \end{pmatrix}\begin{pmatrix} X \\ Y \end{pmatrix}, \qquad [9.14]$$

which can be checked directly. Differentiating this relation, we have:

$$\begin{aligned} \frac{d\Phi}{dx} &= \frac{1}{2}\frac{d}{dx}(P_0Y + X) \\ &= \frac{1}{2}(P_0X - \Delta_y Y) = \frac{1}{2}[P_0(\Phi - P_0\Psi) - \Delta_y(P_0^{-1}\Phi + \Psi)] = P_0\Phi, \end{aligned} \qquad [9.15]$$

and:

$$\begin{aligned} \frac{d\Psi}{dx} &= \frac{1}{2}\left(\frac{dY}{dx} - P_0^{-1}\frac{dX}{dx}\right) = \frac{1}{2}[X - P_0^{-1}(-\Delta_y Y)] = \frac{1}{2}(X - P_0Y) \\ &= \frac{1}{2}[\Phi - P_0\Psi - P_0(P_0^{-1}\Phi + \Psi)] = -P_0\Psi. \end{aligned} \qquad [9.16]$$

For Φ and Ψ, the conditions at $x = 0$ and $x = a$ become:

$$\Phi(0) - P_0\Psi(0) = 0, \quad P_0^{-1}\Phi(a) + \Psi(a) = I. \qquad [9.17]$$

□

We can therefore explicitly state the solutions Φ and Ψ of system [9.13]. Let $S(x), x \geq 0$ be the semi-group with infinitesimal generator P_0. We have:

$$\begin{cases} \Phi(x) = S(a - x)\Phi(a), \\ \Psi(x) = S(x)\Psi(0). \end{cases} \qquad [9.18]$$

Substituting these solutions into the boundary conditions [9.17]:

$$S(a)\Phi(a) - P_0\Psi(0) = 0,$$
$$P_0^{-1}\Phi(a) + S(a)\Psi(0) = I.$$

Hence, by elimination, noting that $I + S(a)^2$ is invertible and that $S(x)$ and P_0 commute:

$$\Phi(a) = (I + S(a)^2)^{-1}P_0,$$
$$\Psi(0) = S(a)(I + S(a)^2)^{-1},$$

and so:

$$\Phi(x) = S(a - x)(I + S(a)^2)^{-1}P_0, \qquad [9.19]$$
$$\Psi(x) = S(a + x)(I + S(a)^2)^{-1}. \qquad [9.20]$$

9.3. Homographic representation of $P(x)$

We have the following homographic representation formula for $P(x)$.

THEOREM 9.2.– The operator P is explicitly given by the formula:

$$P(x) = P_0(I - S(2x))(I + S(2x))^{-1}. \qquad [9.21]$$

PROOF.– Let:

$$W(x) = \Phi(x)\Psi(x)^{-1} \in \mathcal{L}(H_{00}^{1/2}(\mathcal{O}), H_{00}^{1/2}(\mathcal{O})'). \qquad [9.22]$$

Let us now express P as a function of W. We have:

$$\begin{aligned} P(x) &= X(x)Y^{-1}(x) = (\Phi(x) - P_0\Psi(x))(P_0^{-1}\Phi(x) + \Psi(x))^{-1} \\ &= (\Phi(x)\Psi^{-1}(x) - P_0)\Psi(x)((P_0^{-1}\Phi(x)\Psi^{-1}(x) + I)\Psi(x))^{-1} \\ &= (\Phi(x)\Psi^{-1}(x) - P_0)(P_0^{-1}\Phi(x)\Psi^{-1}(x) + I)^{-1} \\ &= (W(x) - P_0)(P_0^{-1}W(x) + I)^{-1}. \end{aligned} \qquad [9.23]$$

Thus, P can be written as a function of W by means of an invertible homography:

$$\begin{aligned} P(x)P_0^{-1}W(x) + P(x) &= W(x) - P_0, \\ W(x) &= (P(x) + P_0)(I - P(x)P_0^{-1})^{-1}. \end{aligned}$$

Denoting by $\mathcal{H}$ the homography in the last line of the above formula, we then have:

$$W(x) = \mathcal{H}(P(x)), \quad P(x) = \mathcal{H}^{-1}(W(x)). \qquad [9.24]$$

Formally taking the derivative w.r.t. x of [9.22], we have:

$$\frac{\mathrm{d}W}{\mathrm{d}x}(x) = \frac{\mathrm{d}\Phi}{\mathrm{d}x}(x)\Psi(x)^{-1} - \Phi(x)\Psi(x)^{-1}\frac{\mathrm{d}\Psi}{\mathrm{d}x}(x)\Psi(x)^{-1}, \qquad [9.25]$$

and from [9.13] it follows that:

$$\frac{\mathrm{d}W}{\mathrm{d}x} = P_0W + WP_0. \qquad [9.26]$$

Note that we know how to calculate $W(0)$, but equation [9.26] is well-posed in the direction of decreasing x and we therefore argue from

$x = a$. We have the following configuration:

$$\begin{array}{ccc} P(x) & \longleftarrow & P(a) \\ \uparrow \mathcal{H}^{-1} & & \downarrow \mathcal{H} \\ W(x) & \longleftarrow & W(a) \end{array} \qquad [9.27]$$

We set:

$$W'(x) = S(2a)W(x). \qquad [9.28]$$

Left-multiplying [9.26] by $S(2a)$ and using the commutative properties of P_0 and $S(x)$, we have:

$$\frac{\mathrm{d}W'}{\mathrm{d}x} = P_0 W' + W' P_0. \qquad [9.29]$$

Furthermore, we can calculate the initial condition for [9.29] at $x = a$ from [9.19], [9.20]:

$$W'(a) = P_0. \qquad [9.30]$$

The solution to this linear equation [9.29] is:

$$W'(x) = P_0 S(2a - 2x), \qquad [9.31]$$

and hence, since the operator $S(x)$ is injective for all $x \geq 0$:

$$W(0) = P_0. \qquad [9.32]$$

From [9.23], left-multiplying by $S(2a)$, we have:

$$P(x)(P_0^{-1}W'(x) + S(2a)) = W'(x) - S(2a)P_0, \qquad [9.33]$$

and so by [9.31] and the semi-group property of $S(x)$, it holds that:

$$P(x)S(2a - 2x)(I + S(2x)) = P_0 S(2a - 2x)(I - S(2x)), \qquad [9.34]$$

from which we deduce [9.21]. □

9.4. Factorization of problem $(\mathcal{P}_0)$ with a Dirichlet condition at $x = 0$

Here, we will further examine the difficulty mentioned in section 3.2.2 that arises when defining an initial condition for the Riccati equation in order to factorize problem $(\mathcal{P}_0)$ if the condition on Γ_0 is of Dirichlet type. We remarked earlier that:

$$\lim_{x \to 0} P(x)^{-1} = 0. \qquad [9.35]$$

In the case studied here, the previous calculations still hold if the condition $X(0) = 0$ is replaced by $Y(0) = 0$. We therefore have:

$$P_0^{-1} S(a)\Phi(a) + \Psi(0) = 0. \qquad [9.36]$$

Expressions [9.19] and [9.20] become:

$$\Phi(x) = S(a - x)(I - S(a)^2)^{-1} P_0, \qquad [9.37]$$

$$\Psi(x) = -S(a + x)(I - S(a)^2)^{-1}. \qquad [9.38]$$

Equation [9.29] for $W'(x) = S(2a)W(x)$ still holds with the condition:

$$W'(a) = -P_0, \qquad [9.39]$$

at $x = a$. As for [9.32], we obtain the initial condition for W at $x = 0$:

$$W(0) = -P_0. \qquad [9.40]$$

We also deduce a homographic representation of $P(x)$ in the form:

$$P(x) = -P_0(I + S(2x))(I - S(2x))^{-1}. \qquad [9.41]$$

Since $S(2x) \to I$ as $x \to 0$, the singularity in P at $x = 0$ is clearly visible in this formulation.

REMARK 9.1.– Note that the homography $\mathcal{H}^{-1}$ defined in [9.24] still holds, and transforms $W(x)$ into $P(x)$. The function $W(x)$ still satisfies [9.26], but now, the initial condition for W corresponds to a singularity of this transformation, which is consistent with [9.35].

Appendix

Gaussian LU Factorization as a Method of Invariant Embedding

In this appendix, we will show that the classical method of factorization of a non-singular $n \times n$ matrix A into the product of a lower triangular matrix L with units along the diagonal and an upper triangular matrix U is a direct consequence of the principle of invariant embedding. Bellman [BEL 57b] showed that the technique of dynamic programming gives an algorithm for solving linear systems with Jacobi-type matrices, that is, tridiagonal symmetric positive definite. Sherman Lehman [SHE 62] subsequently remarked that this algorithm is essentially just the Gaussian elimination . We will show that this is true for all non-singular matrices that can be factorized without pivoting. Note that, in this context, the classical necessary and sufficient condition for matrices to be factorizable without pivoting, namely that all principal submatrices must be non-singular, appears completely natural.

A.1. Invariant embedding for a linear system

Let us consider the linear system:

$$Ax = f, \quad x, f \in \mathbb{R}^n. \tag{A.1}$$

We denote by $[x]_k$ and $[A]_k$ the vector of the first k coordinates of x and the corresponding principal submatrix of order k, respectively. We call $\overline{[A]}_k$ the $k\times$

$(n-k)$ matrix with the $n-k$ columns of A complementary to $[A]_k$ and limited to their first k components and $\overline{[x]}_k$ and the complementary components of $[x]_k$ in x. We call $\underline{[A]}_k$ and $\underline{\overline{[A]}}_k$ the complementary row blocks of $[A]_k$ and $\overline{[A]}_k$, respectively, in A, that is:

$$Ax = \begin{bmatrix} [A]_k & \overline{[A]}_k \\ \underline{[A]}_k & \underline{\overline{[A]}}_k \end{bmatrix} \begin{bmatrix} [x]_k \\ \overline{[x]}_k \end{bmatrix}. \qquad \text{[A.2]}$$

We therefore assume that $[A]_k$ is non-singular for all k. We apply the invariant embedding technique by embedding problem [A.1] into the family:

$$[A]_k[y]_k = [b]_k \quad \text{in } \mathbb{R}^k. \qquad \text{[A.3]}$$

Here, $[b]_k$ is an arbitrary vector of $\mathbb{R}^k$. If we choose $[b]_k = [f]_k - \overline{[A]}_k\overline{[x]}_k$, where $\overline{[x]}_k$ follows from the solution of [A.1], then the solution of [A.3] is the restriction to the first k rows of the solution of [A.1].

Let us consider the problem:

$$[A]_k[x]_k = [f]_k - \overline{[A]}_k\overline{[x]}_k \quad \text{in } \mathbb{R}^k, \qquad \text{[A.4]}$$

where $\overline{[x]}_k$ is arbitrary. According to the hypothesis, this problem has a unique solution that depends affinely on $\overline{[x]}_k$ and therefore on its k-th component. Consequently, there exists β_k dependent on $[f]_k$ and $u_{k,j}, j = k+1, \dots, n$ dependent on $[A]_k$ such that:

$$x_k + \sum_{j=k+1}^{n} u_{k,j}x_j = \beta_k, \qquad \text{[A.5]}$$

where the $x_j, j = k+1, \dots, n$ are arbitrary given variables. (Here, we use the convention that $\sum_{k'}^{k} = 0$ if $k' > k$, so [A.5] gives, in particular, $x_n = \beta_n$ for $k = n$.) As the same reasoning applies to all the components $x_l, l = 1, \dots, k$, we have numbers $\beta_k^l, 1 \le l \le k \le n$ and $u_{k,j}^l, 1 \le l \le k < j \le n$:

$$x_l + \sum_{j=k+1}^{n} u_{k,j}^l x_j = \beta_k^l, \quad l = 1, \dots, k. \qquad \text{[A.6]}$$

For $l = k$, we have:

$$\beta_k^k = \beta_k, \quad u_{k,j}^k = u_{k,j}.$$

Now, let us substitute the values of $[x]_k$ given by [A.5] and [A.6] into the k-th row of [A.4]. Using the fact that $\overline{[x]}_k$ is arbitrary, for the independent term of $\overline{[x]}_k$ we obtain:

$$\sum_{l=1}^{k} a_{k,l}\beta_k^l = f_k, \quad k = 1, \ldots, n, \qquad \text{[A.7]}$$

and for each component of $\overline{[x]}_k$:

$$\sum_{l=1}^{k} a_{k,l}u_{k,j}^l = a_{k,j}, \quad k = 1, \ldots, n-1, \quad j = k+1, \ldots, n. \qquad \text{[A.8]}$$

Likewise, and more generally, by substituting the values of $[x]_k$ given by [A.5] and [A.6] into every row of [A.4], we obtain:

$$\sum_{l=1}^{k} a_{i,l}x_l + \sum_{l=1}^{k}\sum_{j=k+1}^{n} a_{i,l}u_{k,j}^l x_j = \sum_{l=1}^{k} a_{i,l}\beta_k^l, \qquad \text{[A.9]}$$

and consequently:

$$\sum_{l=1}^{k} a_{i,l}\beta_k^l = f_i, \qquad i = 1, \ldots, k, \quad k = 1, \ldots, n, \qquad \text{[A.10]}$$

$$\sum_{l=1}^{k} a_{i,l}u_{k,j}^l = a_{i,j}, \qquad i = 1, \ldots, k, \quad k = 1, \ldots, n-1,$$

$$j = k+1, \ldots, n. \qquad \text{[A.11]}$$

REMARK A.1.– We can therefore interpret $u_{k,j}^l$ as the $(l, j-k)$-th element of the $k \times (n-k)$ matrix $[A]_k^{-1}\overline{[A]}_k$.

Let us now show that β_k^l and $u_{k,j}^l$ are linearly expressed *in the same way* as functions of $\beta_i, i = l, \ldots, k$ and $u_{i,j}, i = l, \ldots, k$, respectively. To define a recurrence relation on k for fixed l, we rewrite the identity [A.6] for steps k and $k-1$, respectively:

$$x_l = -\sum_{j=k+1}^{n} u^l_{k,j} x_j + \beta^l_k = -\sum_{j=k}^{n} u^l_{k-1,j} x_j + \beta^l_{k-1}. \qquad \text{[A.12]}$$

As put forward above, we will take $\overline{[x]}_k$ to be arbitrary. In the second part of [A.12], x_k is a function of $\overline{[x]}_k$ given by [A.5]. Substituting by its value and using the fact that $\overline{[x]}_k$ is arbitrary, we have:

$$u^l_{k,j} = -u^l_{k-1,k} u_{k,j} + u^l_{k-1,j}, \quad 1 \le l \le k < j \le n, \qquad \text{[A.13]}$$

$$\beta^l_k = -u^l_{k-1,k} \beta_k + \beta^l_{k-1}, \quad 1 \le l \le k \le n. \qquad \text{[A.14]}$$

Reiterating this computation up to $l+1$ and summing up gives:

$$u^l_{k,j} = u_{l,j} - \textstyle\sum_{i=l+1}^{k} u^l_{i-1,i} u_{i,j}, \qquad \text{[A.15]}$$

$$\beta^l_k = \beta_l - \textstyle\sum_{i=l+1}^{k} u^l_{i-1,i} \beta_i. \qquad \text{[A.16]}$$

Substituting these values into [A.8] and [A.7] shows the existence of coefficients $\ell_{k,i}, 1 \le i \le k \le n$, for which:

$$\textstyle\sum_{i=1}^{k} \ell_{k,i} u_{i,j} = a_{k,j}, \quad 1 \le i \le k < j \le n, \qquad \text{[A.17]}$$

$$\textstyle\sum_{i=1}^{k} \ell_{k,i} \beta_i = f_k, \quad 1 \le i \le k \le n. \qquad \text{[A.18]}$$

Let us elucidate this computation, for example, on [A.7]. We have:

$$\begin{aligned}
\sum_{l=1}^{k} a_{k,l} \beta^l_k &= \sum_{l=1}^{k} a_{k,l} \left(\beta_l - \sum_{i=l+1}^{k} u^l_{i-1,i} \beta_i \right) \\
&= -\sum_{l=1}^{k} a_{k,l} \left(\sum_{i=l}^{k} u^l_{i-1,i} \beta_i \right) \\
&= -\sum_{i=1}^{k} \beta_i \left(\sum_{l=1}^{i} a_{k,l} u^l_{i-1,i} \right),
\end{aligned}$$

where we have set $u_{l-1,l}^l = -1$. Therefore, defining $\ell_{k,i}$ for $1 \leq i \leq k \leq n$ by:

$$\ell_{k,i} = -\sum_{m=1}^{i} a_{k,m} u_{i-1,i}^m = a_{k,i} - \sum_{m=1}^{i-1} a_{k,m} u_{i-1,i}^m, \qquad [A.19]$$

relation [A.18] is satisfied. Likewise, [A.17] is satisfied. Because of [A.15], we have for $1 \leq i \leq k \leq n$:

$$\ell_{k,i} = a_{k,i} - \sum_{m=1}^{i-1} a_{k,m} \sum_{j=m}^{i-1} u_{j-1,j}^m u_{j,i} \qquad [A.20]$$

$$= a_{k,i} - \sum_{j=1}^{i-1} u_{j,i} \sum_{m=1}^{j} a_{k,m} u_{j-1,j}^m \qquad [A.21]$$

$$= a_{k,i} - \sum_{j=1}^{i-1} u_{j,i} \ell_{k,j}. \qquad [A.22]$$

If we extend u by $u_{i,i} = 1$ and $u_{i,j} = 0$ for $j < i$, and ℓ by $\ell_{i,j} = 0$ for $j \geq i$, then by combining [A.17] and [A.22] we obtain:

$$\sum_{i=1}^{n} \ell_{k,i} u_{i,j} = a_{k,j}, \quad k = 1, \ldots, n \quad j = 1, \ldots, n, \qquad [A.23]$$

or in matrix form:

$$A = LU. \qquad [A.24]$$

We have obtained the factorization of A into the product of a lower triangular matrix L, with elements $\ell_{i,j}$, and an upper triangular matrix U with units on the diagonal, with elements $u_{i,j}$ (Crout's form) with only the hypothesis that the principal submatrices $[A]_k$ are non-singular. Relation [A.18] gives a triangular system for the downward phase and [A.5] for the upward phase. Note that to obtain the factorization [A.23], we have used the result [A.17] concerning the lower triangular part of A that follows from the definition of L and the result [A.22], which in turn follows from an application of the invariant embedding principle and in particular from the fact that we assumed "the present state" of the system, that is, the variable $\overline{[x]}_k$ to be arbitrary. This presentation of the Gaussian factorization gives the same results L and U as the Crout method because we know that this factorization LU is unique if we require U to have units on the diagonal.

REMARK A.2.– Requiring relation [A.5] to take this form (coefficient of x_k is one) leads to Crout's factorization. With a different coefficient $u_{k,k}$, it is possible to show that choosing $\ell_{k,k} = 1$ would lead to the classical Gaussian factorization.

Classically, we compute L and U by identification from formula [A.23]: at the stage k, we compute the k-th column of L using the fact that $u_{k,k} = 1$ by identification with the k-th column of A and then the k-th row of U using the k-th row of A. Let us show that we can make an equivalent computation using the coefficients $u^l_{k-1,k}$. Indeed, suppose that $u^l_{m-1,m}$ are known for $1 \leq l \leq m \leq k$ and $u_{i-1,j}$ are known for $j \geq i-1$. We can compute $\ell_{k,i}$ for $1 \leq i \leq k$ using the definition [A.19]. Now, we can obtain $u_{k,j}$ for $j \geq k$ using [A.17]:

$$u_{k,j} = \frac{1}{\ell_{k,k}}(a_{k,j} - \sum_{i=1}^{k-1} \ell_{k,i} u_{i,j}). \qquad \text{[A.25]}$$

Finally, for the recurrence on k to be valid, we compute $u^m_{k,k+1}$ for $1 \leq m \leq k+1$ by decreasing recurrence on m starting from $u^{k+1}_{k,k+1} = -1$. By [A.15], we have:

$$u^m_{k,k+1} = u_{m,k+1} - \sum_{i=m+1}^{k} u^m_{i-1,i} u_{i,k+1}. \qquad \text{[A.26]}$$

With this formulation, we can clearly see, at least for L, the well-known rule of profile conservation: by formula [A.19], it is obvious that if $a_{k,m} = 0$ for $1 \leq m \leq i$, this also applies for $\ell_{k,i}$.

Let us now consider the following situation: we assume that, at the stage k, we have to solve [A.1] several times for different functions $\overline{[f]}_k$ and then different matrices $\underline{\overline{[A]}}_k$, despite only being interested in the $\overline{[x]}_k$ part of the unknown. We must therefore eliminate the $[x]_k$ component, which is classically done using the Schur complement formula:

$$(\underline{\overline{[A]}}_k - \underline{[A]}_k [A]_k^{-1} \overline{[A]}_k)\overline{[x]}_k = \overline{[f]}_k - \underline{[A]}_k [A]_k^{-1} [f]_k. \qquad \text{[A.27]}$$

We will show that the matrix $\underline{[A]}_k[A]_k^{-1}\overline{[A]}_k$ has a simple expression in terms of the first k columns of L and the first k rows of U. We call $g_{k,j}^i$ the element (i,j) of $\underline{[A]}_k[A]_k^{-1}\overline{[A]}_k$. From remark A.1, we have:

$$g_{k,j}^i = \sum_{m=1}^{k} a_{i,m}u_{k,j}^m, \quad \text{for } i,j > k. \tag{A.28}$$

Using [A.15]:

$$\begin{aligned} g_{k,j}^i &= \sum_{m=1}^{k} a_{i,m}u_{m,j} - \sum_{m=1}^{k} a_{i,m} \sum_{l=m+1}^{k} u_{l-1,l}^m u_{l,j} \\ &= \sum_{m=1}^{k} a_{i,m}u_{m,j} - \sum_{l=2}^{k} u_{l,j} \sum_{m=1}^{l-1} a_{i,m}u_{j-1,l}^m. \end{aligned}$$

By [A.19]:

$$g_{k,j}^i = \sum_{m=1}^{k} a_{i,m}u_{m,j} + \sum_{m=2}^{k} u_{m,j}(\ell_{i,m} - a_{i,m}) \tag{A.29}$$

$$= a_{i,1}u_{1,j} + \sum_{m=2}^{k} \ell_{i,m}u_{m,j} \tag{A.30}$$

$$g_{k,j}^i = \sum_{m=1}^{k} \ell_{i,m}u_{m,j}. \tag{A.31}$$

A.2. Block tridiagonal systems

Let us consider the case of a system whose matrix is defined in blocks by:

$$A = \begin{bmatrix} A_1 & C_1 & & \\ B_2 & A_2 & C_2 & 0 \\ 0 & \ddots & \ddots & \ddots \\ & & B_n & A_n \end{bmatrix}, \tag{A.32}$$

where A_i is $n_i \times n_i$, B_i is $n_i \times n_{i-1}$, and C_i is $n_i \times n_{i+1}$. We denote x_i the i-th subvector of length n_i extracted from x. The family of problems defined by [A.4] only depends on x_{k+1} at order k:

$$[A]_k[x]_k = [f]_k - \begin{bmatrix} 0 \\ \cdot \\ 0 \\ C_k \end{bmatrix} x_{k+1}, \qquad [A.33]$$

for an arbitrary given x_{k+1}. The identity [A.5] shows that there exists an $n_k \times n_{k+1}$ matrix U_k, such that:

$$x_k + U_k x_{k+1} = \beta_k. \qquad [A.34]$$

As for [A.6], we obtain the $n_{k-1} \times n_{k+1}$ matrix V_k:

$$x_{k-1} + V_k x_{k+1} = \gamma_k. \qquad [A.35]$$

By identification after substituting these values into [A.33], using the fact that x_k is arbitrary:

$$A_1 U_1 = C_1, \quad B_k V_k + A_k U_k = C_k,$$
$$A_1 \beta_1 = f_1, \quad B_k \gamma_k + A_k \beta_k = f_k.$$

By a computation similar to [A.12], we obtain:

$$V_k = -U_{k-1} U_k,$$
$$\gamma_k = \beta_{k-1} - U_{k-1} \beta_k.$$

Setting:

$$L_1 = A_1, \qquad L_k = A_k - B_k U_{k-1},$$

we have:

$$A_k = L_k + B_k U_{k-1}, \qquad C_k = L_k U_k.$$

We then have the factorization:

$$[A]_k = \begin{bmatrix} L_1 & & & \\ B_2 & L_2 & & 0 \\ 0 & \ddots & \ddots & \\ & & B_k & L_k \end{bmatrix} \begin{bmatrix} I & U_1 & & \\ & \ddots & \ddots & 0 \\ 0 & & I & U_{k-1} \\ & & 0 & I \end{bmatrix}. \qquad [A.36]$$

Therefore, L_k is non-singular. Indeed, if not, the upper triangular block matrix of [A.36] and therefore $[A]_k$ would be singular because the upper triangular matrix, with I on its diagonal, is non-singular. This would contradict our hypothesis. We can therefore compute U_k for decreasing k by:

$$U_1 = A_1^{-1} C_1,$$
$$U_k = L_k^{-1} C_k \quad \text{for } k = 2, \ldots, n-1.$$

Likewise, for β_k:

$$\beta_1 = A_1^{-1} f_1,$$
$$\beta_k = L_k^{-1}(f_k - B_k \beta_{k-1}) \quad \text{for } k = 2, \ldots, n-1.$$

Finally, the solution x_k for decreasing k is given by:

$$x_n = \beta_n,$$
$$x_k = \beta_k - U_k x_{k+1} \quad \text{for } k = n-1, \ldots, 1.$$

We have found Thomas's algorithm or at least a modification assigning the units on the diagonal to the upper triangular half of the matrix. The factorization we have obtained is written as:

$$A = \begin{bmatrix} L_1 & & & \\ B_2 & L_2 & & 0 \\ 0 & \ddots & \ddots & \\ & & B_n & L_n \end{bmatrix} \begin{bmatrix} I & U_1 & & \\ & I & U_2 & 0 \\ 0 & & \ddots & \ddots \\ & & 0 & I \end{bmatrix}. \qquad \text{[A.37]}$$

REMARK A.3.– We point out the analogy with the computation of the optimal feedback for a quadratic linear system with discrete time: if A_k represents the dynamics of the controlled system, $L_k = A_k - B_k U_{k-1}$ is the dynamics of a system looped by the optimal feedback.

In this framework, the Schur complement is given by the application of formula [A.31] by:

$$\overline{\underline{[A]}}_k - B_{k+1} U_k. \qquad \text{[A.38]}$$

A.2.1. *Backward invariant embedding*

We will now present the formulas obtained by invariant embedding for decreasing k starting from n. This computation could also be performed for general systems, but we will limit ourselves to studying cases that are block tridiagonal. Therefore, we will consider the sub-problem relative to the last k variables $\overline{[x]}_k$:

$$\overline{\underline{[A]}}_k \overline{[x]}_k = \overline{[f]}_k - \begin{bmatrix} B_{k+1} \\ 0 \\ \cdot \\ 0 \end{bmatrix} x_k, \qquad \text{[A.39]}$$

for arbitrary x_k. We have the identity:

$$x_{k+1} + U'_{k+1} x_k = \beta'_{k+1}, \qquad \text{[A.40]}$$

and x_{k+2} can also be expressed as a function of x_k:

$$x_{k+2} + V'_{k+1}x_k = \gamma'_{k+1}. \qquad [A.41]$$

Substituting into [A.39]:

$$A_k U'_k + C_k V'_k = B_k. \qquad [A.42]$$

Using a substitution computation similar to that performed earlier, we find:

$$V'_k = -U'_{k+1}U'_k. \qquad [A.43]$$

Setting:

$$L'_k = A_k - C_k U'_{k+1}, \qquad [A.44]$$

we can compute U'_k and L'_k for decreasing k, starting from:

$$U'_n = A_n^{-1}B_n, \qquad L'_n = A_n.$$

The factorization thus obtained can be written as:

$$A = \begin{bmatrix} L'_1 & C_1 & & \\ & L'_2 & C_2 & 0 \\ 0 & & \ddots & \ddots \\ & & & L'_n \end{bmatrix} \begin{bmatrix} I & & & \\ U'_2 & I & & 0 \\ 0 & \ddots & \ddots & \\ & & U'_n & I \end{bmatrix}. \qquad [A.45]$$

This is the product of an upper triangular matrix and a lower triangular matrix.

REMARK A.4.– There are no "local" relations, that is, for neighboring indices, between U_k and L_k in the factorization [A.37] and the U'_k and L'_k obtained here, because in the first case they correspond to the upper left part of A, whereas here they are dependent on the lower right part.

Bibliography

[ABO 03] ABOU-KANDIL H., FREILING G., IONESCU V., *et al.*, *Matrix Riccati Equations in Control and Systems Theory*, Birkhäuser, 2003.

[ANG 68a] ANGEL E., "Discrete invariant imbedding and elliptic boundary-value problems over irregular regions", *Journal of Mathematical Analysis and Applications*, vol. 23, pp. 471–484, 1968.

[ANG 68b] ANGEL E., "Dynamic programming and linear partial differential equations", *Journal of Mathematical Analysis and Applications*, vol. 23, pp. 628–638, 1968.

[ANG 69] ANGEL E., "A building block technique for boundary value problems over irregular regions", *Journal of Mathematical Analysis and Applications*, vol. 26, pp. 75–81, 1969.

[ANG 70a] ANGEL E., "Invariant imbedding, difference equations, and elliptic boundary value problems", *Journal of Computer and System Sciences*, vol. 4, pp. 473–491, 1970.

[ANG 70b] ANGEL E., KALABA R.E., "A one-sweep numerical method for vector matrix difference equations with two point boundary conditions", *Journal of Optimization Theory and Applications*, vol. 6, no. 5, pp. 345–355, 1970.

[ANG 71a] ANGEL E., BELLMAN R., *Dynamic Programming and Partial Differential Equations*, Academic Press, 1971.

[ANG 71b] ANGEL E., JAIN A.K., "Initial-value transformations for elliptic boundary value problems", *Journal of Mathematical Analysis and Applications*, vol. 35, pp. 496–502, 1971.

[ANG 86] ANGEL E., "From dynamic programming to fast transforms", *Journal of Mathematical Analysis and Applications*, vol. 119, pp. 82–89, 1986.

[ANG 94] ANGEL E., LEONG P., "Multidimensional dynamic programming on massively parallel computers", *Computers & Mathematics with Applications*, vol. 27, nos. 9–10, pp. 67–75, 1994.

[ÅST 84] ÅSTRÖM K.J., WITTENMARK B., *Computer-Controlled Systems: Theory and Design*, Prentice-Hall, 1984.

[BEL 57a] BELLMAN R., *Dynamic Programming*, Princeton University Press, 1957.

[BEL 57b] BELLMAN R., "On some applications of dynamic programming to matrix theory", *Illinois Journal of Mathematics*, vol. 1, pp. 297–301, 1957.

[BEN 71] BENSOUSSAN A., *Filtrage optimal des systèmes linéaires*, Dunod, 1971.

[BEN 07] BENSOUSSAN A., DA PRATO G., DELFOUR M., *et al.*, *Representation and Control of Infinite Dimensional Systems*, Birkhäuser, 2007.

[BEN 11] BEN ABDA A., HENRY J., JDAY F., "Boundary data completion: the method of boundary value problem factorization", *Inverse Problems*, vol. 27, no. 5, Page055014, 2011.

[BOU 07] BOUARROUDJ N., HENRY J., LOURO B., *et al.*, "On a direct study of an operator Riccati equation appearing in boundary value problems factorization", *Comptes Rendus de l'Académie des Sciences – Series I – Mathematics*, 2007.

[BOU 08] BOUARROUDJ N., HENRY J., LOURO B., *et al.*, "On a direct study of an operator Riccati equation appearing in boundary value problems factorization", *Applied Mathematical Sciences*, vol. 2, no. 46, pp. 2247–2257, 2008.

[BOY 06] BOYER F., FABRIE P., "Eléments d'analyse pour l'étude de quelques modèles d'écoulements de fluides visqueux incompressibles", *Mathématiques & Applications*, vol. 52, 2006.

[BRE 99] BREZIS H., *Analyse fonctionnelle: théorie et applications*, Dunod, Paris, 1999.

[CHA 04] CHAMPAGNE I., Méthodes de Factorisation des Equations aux Dérivées Partielles, Thesis, Ecole Polytechnique, 2004.

[CIA 98] CIARLET P.G., *Introduction à l'analyse numérique matricielle et à l'optimisation*, Masson, 1998.

[COL 71] COLLINS D., ANGEL E., "The diagonal decomposition technique applied to the dynamic programming solution of elliptic partial differential equations", *Journal of Mathematical Analysis and Applications*, vol. 33, pp. 467–481, 1971.

[DEL 01] DELFOUR M., ZOLÉSIO J.P., *Shapes and Geomtries: Analysis, Differential Calculus and Optimization*, SIAM, 2001.

[HAD 10] HADAMARD J., *Leçons sur le calcul des variations*, A. Hermann et Fils, Paris, 1910.

[HEN 96] HENRY J., YVON J.P., "On the use of space invariant embedding to solve optimal control problems for second order elliptic equations", in DOLEŽAL J., FIDLER J., (eds.), *System Modelling and Optimization*, Springer-Verlag, 1996.

[HEN 04] HENRY J., RAMOS A.M., "Factorization of second order elliptic boundary value problems by dynamic programming", *Nonlinear Analysis: Theory, Methods & Applications*, vol. 59, no. 5, pp. 629–647, 2004.

[HEN 07] HENRY J., "The computing zoom", in GLOWINSKI R., ZOLÉSIO J.P., (eds.), *Free and Moving Boundaries: Analysis, Simulation and Control*, Chapman and Hall/CRC, Taylor and Francis Group, 2007.

[HEN 08] HENRY J., RAMOS A.M., "Study of the initial value problems appearing in a method of factorization of second-order elliptic boundary value problems", *Nonlinear Analysis: Theory, Methods & Applications*, vol. 68, no. 10, pp. 2984–3008, 2008.

[HEN 10] HENRY J., LOURO B., OREY M., "Factorization of overdetermined boundary value problems", CAKAJ S., ed., *Modeling, Simulation and Optimization – Tolerance and Optimal Control*, InTech, pp. 207–221, 2010.

[JAI 75] JAIN P.C., KADALBAJOO M.K., "Dynamic programming solutions of mildly non-linear elliptic problems over irregular regions", *International Journal of Computer Mathematics*, vol. 5, pp. 231–239, 1975.

[JDA 12] JDAY F., La méthode de factorisation des problèmes aux limites: Application à la reconstruction de données frontières, Thesis, ENIT, Tunis, 2012.

[KAT 75] KATO T., "Quasi-linear equations of evolution, with applications to partial differential equations", in DOLD A., ECKMANN B., (eds.), *Spectral Theory and Differential Equations*, Springer-Verlag, 1975.

[LIO 68a] LIONS J.L., *Contrôle Optimal de Systèmes Gouvernés par des Équations aux Dérivées Partielles*, Dunod, 1968.

[LIO 68b] LIONS J.L., MAGENES E., *Problèmes aux Limites Non Homogènes et Applications*, vol. 1, Dunod, 1968.

[LIO 68c] LIONS J.L., MAGENES E., *Problèmes aux Limites Non Homogènes et Applications*, vol. 2, Dunod, 1968.

[LIO 73] LIONS J.L., *Perturbations singulières dans les problèmes aux limites et en contrôle optimal*, Springer-Verlag, Berlin, 1973.

[LIO 83] LIONS J.L., *Contrôle des systèmes distribués singuliers*, Gauthier-Villars, Paris, 1983.

[MAL 77] MALVERN L.E., *Introduction to the Mechanics of a Continuous Medium*, Prentice Hall, 1977.

[ORE 12] OREY M., Factorization of elliptic boundary value problems by invariant embedding and application to overdetermined problems, PhD Thesis, Universidade Nova de Lisboa, 2012.

[PAZ 83] PAZY A., *Semigroups of Linear Operators and Applications to Partial Differential Equations*, Springer-Verlag, 1983.

[QUA 99] QUARTERONI A., VALLI A., *Domain Decomposition Methods for Partial Differential Equations*, Oxford University Press, 1999.

[RAM 96] RAMOS A.M., Algunos problemas en ecuaciones en derivadas partiales relacionados con la teoría de Control, PhD Thesis, Universidad Computense de Madrid, 3 July 1996.

[RAM 12] RAMOS A.M., *Introducción al análisis matemático del método de elementos finitos*, Editorial Complutense, 2012.

[SCH 59] SCHWARTZ L., "Théorie des distributions à valeurs vectorielles, I, II", *Annales Institut Fourier*, vol. I, no. 7, pp. 1–141, 1959.

[SHE 62] SHERMAN LEHMAN R., "Dynamic programming and Gaussian elimination", *Journal of Mathematical Analysis and Applications*, vol. 5, pp. 499–501, 1962.

[SOA 06] SOARES M.D.C. Factorization by invariant embedding of elliptic problems: circular and star shaped domains, PhD thesis, Universidade Nova de Lisboa, 2006.

[SOK 92] SOKOLOWSKI J., ZOLESIO J.P., *Introduction to Shape Optimization, Shape Sensitivity Analysis*, Springer, 1992.

[TAR 74] TARTAR L., "Sur l'étude directe d'équations non linéaires intervenant en théorie du contrôle optimal", *Journal of Functional Analysis*, vol. 6, pp. 1–47, 1974.

[TEM 71] TEMAM R., "Sur l'équation de Riccati associée à des opérateurs non bornés, en dimension infinie", *Journal of Functional Analysis*, vol. 7, no. 1, pp. 85–115, 1971.

Index

Kolibri 360 GmbH
Ferdinand-Jühlke-Straße 7
99095 Erfurt, Deutschland
produktsicherheit@kolibri360.de